Philosophic Foundations of Quantum Mechanics

HANS REICHENBACH

DOVER PUBLICATIONS, INC.
Mineola, New York

Copyright

Copyright © 1944 by The Regents of the University of California
Copyright © renewed by Maria Reichenbach
All rights reserved.

Bibliographical Note

This Dover edition, first published in 1998, is an unabridged republication of the paperback edition (1975) of the work originally published in 1944 by the University of California Press, Berkeley and Los Angeles.

Library of Congress Cataloging-in-Publication Data

Reichenbach, Hans, 1891–1953.
    Philosophic foundations of quantum mechanics / by Hans Reichenbach. – Dover ed.
        p.   cm.
    "This Dover edition, first published in 1998, is an unabridged republication of the paperback edition (1975) of the work originally published in 1944 by the University of California Press, Berkeley and Los Angeles."
    Includes bibliographical references and index.
    ISBN-13: 978-0-486-40459-2
    ISBN-10: 0-486-40459-5
    1. Quantum theory. 2. Physics–Philosophy. I. Title.
QC174.12.R45   1998
530.12–dc21   98-40052
                                                            CIP

Manufactured in the United States by LSC Communications
40459510   2018
www.doverpublications.com

PREFACE

Two great theoretical constructions have shaped the face of modern physics: the theory of relativity and the theory of quanta. The first has been, on the whole, the discovery of one man, since the work of Albert Einstein has remained unparalleled by the contributions of others who, like Hendrik Anton Lorentz, came very close to the foundations of special relativity, or, like Hermann Minkowski, determined the geometrical form of the theory. It is different with the theory of quanta. This theory has been developed by the collaboration of a number of men each of whom has contributed an essential part, and each of whom, in his work, has made use of the results of others.

The necessity of such teamwork seems to be deeply rooted in the subject matter of quantum theory. In the first place, the development of this theory has been greatly dependent on the production of observational results and on the exactness of the numerical values obtained. Without the help of the army of experimenters who photographed spectral lines or watched the behavior of elementary particles by means of ingenious devices, it would have been impossible ever to carry through the theory of the quanta even after its foundations had been laid. In the second place, these foundations are very different in logical form from those of the theory of relativity. They have never had the form of one unifying principle, not even after the theory has been completed. They consist of a set of principles which, despite their mathematical elegance, do not possess the suggestive character of a principle which convinces us at first sight, as does the principle of relativity. And, finally, they depart much further from the principles of classical physics than the theory of relativity ever did in its criticism of space and time; their implications include, in addition to a transition from causal laws to probability laws, a revision of philosophical ideas about the existence of unobserved objects, even of the principles of logic, and reach down to the deepest fundamentals of the theory of knowledge.

In the development of the theoretical form of quantum physics, we can distinguish four phases. The first phase is associated with the names of Max Planck, Albert Einstein, and Nils Bohr. Planck's introduction of the quanta in 1900 was followed by Einstein's extension of the quantum concept toward that of a needle radiation (1905). The decisive step, however, was made in Bohr's application (1913) of the quantum idea to the analysis of the structure of the atom, which led to a new world of physical discoveries.

The second phase, which began in 1925, represents the work of a younger generation which had been trained in the physics of Planck, Einstein, and Bohr, and started where the older ones had stopped. It is a most astonishing fact that this phase, which led up to quantum mechanics, began without a clear insight into what was actually being done. Louis de Broglie introduced waves as companions of particles; Erwin Schrödinger, guided by mathematical

PREFACE

analogies with wave optics, discovered the two fundamental differential equations of quantum mechanics; Max Born, Werner Heisenberg, Pascual Jordan, and, independently of this group, Paul A. M. Dirac constructed the matrix mechanics which seemed to defy any wave interpretation. This period represents an amazing triumph of mathematical technique which, masterly applied and guided by a physical instinct more than by logical principles, determined the path to the discovery of a theory which was able to embrace all observable data. All this was done in a very short time; by 1926 the mathematical shape of the new theory had become clear.

The third phase followed immediately; it consisted in the physical interpretation of the results obtained. Schrödinger showed the identity of wave mechanics and matrix mechanics. Born recognized the probability interpretation of the waves. Heisenberg saw that the mathematical mechanism of the theory involves an unsurmountable uncertainty of predictions and a disturbance of the object by the measurement. And here once more Bohr intervened in the work of the younger generation and showed that the description of nature given by the theory was to leave open a specific ambiguity which he formulated in his principle of complementarity.

The fourth phase continues up to the present day; it is filled with constant extensions of the results obtained toward further and further applications, including the application to new experimental results. These extensions are combined with mathematical refinements; in particular, the adaptation of the mathematical method to the postulates of relativity is in the foreground of the investigations. We shall not speak of these problems here, since our inquiry is concerned with the logical foundations of the theory.

It was with the phase of the physical interpretations that the novelty of the logical form of quantum mechanics was realized. Something had been achieved in this new theory which was contrary to traditional concepts of knowledge and reality. It was not easy, however, to *say* what had happened, i.e., to proceed to the *philosophical* interpretation of the theory. Based on the physical interpretations given, a philosophy for common use was developed by the physicists which spoke of the relation of subject and object, of pictures of reality which must remain vague and unsatisfactory, of operationalism which is satisfied when observational predictions are correctly made, and renounces interpretations as unnecessary ballast. Such concepts may appear useful for the purpose of carrying on the merely technical work of the physicist. But it seems to us that the physicist, whenever he tried to be conscious of what he did, could not help feeling a little uneasy with this philosophy. He then became aware that he was walking, so to speak, on the thin ice of a superficially frozen lake, and he realized that he might slip and break through at any moment.

It was this feeling of uneasiness which led the author to attempt a philosophical analysis of the foundations of quantum mechanics. Fully aware that philosophy should not try to construct physical results, nor try to prevent

PREFACE

physicists from finding such results, he nonetheless believed that a logical analysis of physics which did not use vague concepts and unfair excuses was possible. The philosophy of physics should be as neat and clear as physics itself; it should not take refuge in conceptions of speculative philosophy which must appear outmoded in the age of empiricism, nor use the operational form of empiricism as a way to evade problems of the logic of interpretations. Directed by this principle the author has tried in the present book to develop a philosophical interpretation of quantum physics which is free from metaphysics, and yet allows us to consider quantum mechanical results as statements about an atomic world as real as the ordinary physical world.

It scarcely will appear necessary to emphasize that this philosophical analysis is carried through in deepest admiration of the work of the physicists, and that it does not pretend to interfere with the method of physical inquiry. All that is intended in this book is clarification of concepts; nowhere in this presentation, therefore, is any contribution toward the solution of physical problems to be expected. Whereas physics consists in the analysis of the physical world, philosophy consists in the analysis of our knowledge of the physical world. The present book is meant to be philosophical in this sense.

The division of the book is so planned that the first part presents the general ideas on which quantum mechanics is based; this part, therefore, outlines our philosophical interpretation and summarizes its results. The presentation is such that it does not presuppose mathematical knowledge, nor an acquaintance with the methods of quantum physics. In the second part we present the outlines of the mathematical methods of quantum mechanics; this is so written that a knowledge of the calculus should enable the reader to understand the exposition. Since we possess today a number of excellent textbooks on quantum mechanics, such an exposition may appear unnecessary; we give it, however, in order to open a short cut toward the mathematical foundations of quantum mechanics for all those who do not have the time for thorough studies of the subject, or who would like to see in a short review the methods which they have applied in many individual problems. Our presentation, of course, makes no claim to be complete. The third part deals with the various interpretations of quantum mechanics; it is here that we make use of both the philosophical ideas of the first part and the mathematical formulations of the second. The properties of the different interpretations are discussed, and an interpretation in terms of a three-valued logic is constructed which appears as a satisfactory logical form of quantum mechanics.

I am greatly indebted to Dr. Valentin Bargmann of the Institute of Advanced Studies in Princeton for his advice in mathematical and physical questions; numerous improvements in the presentation, in Part II in particular, are due to his suggestions. I wish to thank Dr. Norman C. Dalkey of the University of California, Los Angeles, and Dr. Ernest H. Hutten, formerly at Los Angeles, now in the University of Chicago, for the opportunity of discussing with them questions of a logical nature, and for their assistance in matters of style and terminology. Finally I wish to thank the staff of the University of California Press for the care and consideration with which they have edited my book and for their liberality in following my wishes concerning some deviations from established usage in punctuation.

A presentation of the views developed in this book, including an exposition of the system of three-valued logic introduced in § 32, was given by the author at the Unity of Science Meeting in the University of Chicago on September 5, 1941.

HANS REICHENBACH
Department of Philosophy,
University of California,
Los Angeles

June, 1942

CONTENTS

Part I: General Considerations

	PAGE
§ 1. Causal laws and probability laws	1
§ 2. The probability distributions	5
§ 3. The principle of indeterminacy	9
§ 4. The disturbance of the object by the observation	14
§ 5. The determination of unobserved objects	17
§ 6. Waves and corpuscles	20
§ 7. Analysis of an interference experiment	24
§ 8. Exhaustive and restrictive interpretations	32

Part II: Outlines of the Mathematics of Quantum Mechanics

§ 9. Expansion of a function in terms of an orthogonal set	45
§ 10. Geometrical interpretation in the function space	52
§ 11. Reversion and iteration of transformations	58
§ 12. Functions of several variables and the configuration space	64
§ 13. Derivation of Schrödinger's equation from de Broglie's principle	66
§ 14. Operators, eigen-functions, and eigen-values of physical entities	72
§ 15. The commutation rule	76
§ 16. Operator matrices	78
§ 17. Determination of the probability distributions	81
§ 18. Time dependence of the ψ-function	85
§ 19. Transformation to other state functions	90
§ 20. Observational determination of the ψ-function	91
§ 21. Mathematical theory of measurement	95
§ 22. The rules of probability and the disturbance by the measurement	100
§ 23. The nature of probabilities and of statistical assemblages in quantum mechanics	105

Part III: Interpretations

	Page
§ 24. Comparison of classical and quantum mechanical statistics	111
§ 25. The corpuscle interpretation	118
§ 26. The impossibility of a chain structure	122
§ 27. The wave interpretation	129
§ 28. Observational language and quantum mechanical language	136
§ 29. Interpretation by a restricted meaning	139
§ 30. Interpretation through a three-valued logic	144
§ 31. The rules of two-valued logic	148
§ 32. The rules of three-valued logic	150
§ 33. Suppression of causal anomalies through a three-valued logic	160
§ 34. Indeterminacy in the object language	166
§ 35. The limitation of measurability	169
§ 36. Correlated systems	170
§ 37. Conclusion	176
Index	179

Part I
GENERAL CONSIDERATIONS

§ 1. Causal Laws and Probability Laws

The philosophical problems of quantum mechanics are centered around two main issues. The first concerns the transition from causal laws to probability laws; the second concerns the interpretation of unobserved objects. We begin with the discussion of the first issue, and shall enter into the analysis of the second in later sections.

The question of replacing causal laws by statistical laws made its appearance in the history of physics long before the times of the theory of quanta. Since the time of Boltzmann's great discovery which revealed the second principle of thermodynamics to be a statistical instead of a causal law, the opinion has been repeatedly uttered that a similar fate may meet all other physical laws. The idea of determinism, i.e., of strict causal laws governing the elementary phenomena of nature, was recognized as an extrapolation inferred from the causal regularities of the macrocosm. The validity of this extrapolation was questioned as soon as it turned out that macrocosmic regularity is equally compatible with irregularity in the microcosmic domain, since the law of great numbers will transform the probability character of the elementary phenomena into the practical certainty of statistical laws. Observations in the macrocosmic domain will never furnish any evidence for causality of atomic occurrences so long as only effects of great numbers of atomic particles are considered. This was the result of unprejudiced philosophical analysis of the physics of Boltzmann.[1]

With this result a decision of the question was postponed until it was possible to observe macrocosmic effects of individual atomic phenomena. Even with the use of observations of this kind, however, the question is not easily answered, but requires the development of a more profound logical analysis.

Whenever we speak of strictly causal laws we assume them to hold between idealized physical states; and we know that actual physical states never cor-

---

[1] It is scarcely possible to say who was the first to formulate this philosophical idea. We have no published utterances of Boltzmann indicating that he thought of the possibility of abandoning the principle of causality. In the decade preceding the formulation of quantum mechanics the idea was often discussed. F. Exner, in his book, *Vorlesungen über die physikalischen Grundlagen der Naturwissenschaften* (Vienna, 1919), is perhaps the first to have clearly stated the criticism which we gave above: "Let us not forget that the principle of causality and the need for causality has been suggested to us exclusively by experiences with macrocosmic phenomena and that a transference of the principle to microcosmic phenomena, viz. the assumption that every individual occurrence be strictly causally determined, has no longer any justification based on experience."—p. 691. With reference to Exner, E. Schrödinger has expressed similar ideas in his inaugural address in Zurich, 1922, published in *Naturwissenschaften*, 17: 9 (1929).

PART I. GENERAL CONSIDERATIONS

respond exactly to the conditions assumed for the laws. This discrepancy has often been disregarded as irrelevant, as being due to the imperfection of the experimenter and therefore negligible in a statement about causality as a property of nature. With such an attitude, however, the way to a solution of the problem of causality is barred. Statements about the physical world have meaning only so far as they are connected with verifiable results; and a statement about strict causality must be translatable into statements about observable relations if it is to have a utilizable meaning. Following this principle we can interpret the statement of causality in the following way.

If we characterize physical states in observational terms, i.e., in terms of observations as they are actually made, we know that we can construct probability relations between these states. For instance, if we know the inclination of the barrel of a gun, the powder charge, and the weight of the shell, we can predict the point of impact with a certain probability. Let A be the so-defined initial conditions and B a description of the point of impact; then we have a probability implication

$$A \underset{p}{\Rightarrow} B \quad (1)$$

which states that if A is given, B will happen with a determinate probability p. From this empirically verifiable relation we pass to an ideal relation by considering ideal states A' and B' and stating a logical implication

$$A' \supset B' \quad (2)$$

between them, which represents a law of mechanics. Since we know, however, that from the observational state A we can infer only with some probability the existence of the ideal state A', and that similarly we have only a probability relation between B and B', the logical implication (2) cannot be utilized. It derives its physical meaning only from the fact that in all cases of applications to observable phenomena it can be replaced by the probability implication (1). What then is the meaning of a statement saying that if we knew exactly the initial conditions we could predict with certainty the future states resulting from them? Such a statement can be meaningfully said only in the sense of a transition to a limit. Instead of characterizing the initial conditions of shooting only by the mentioned three parameters, the inclination of the barrel, the powder charge, and the weight of the shell, we can consider further parameters, such as the resistance of the air, the rotation of the earth, etc. As a consequence, the predicted value will change; but we know that with such a more precise characterization also the probability of the prediction increases. From experiences of this kind we have inferred that the probability p can be made to approach the value 1 as closely as we want by the introduction of further parameters into the analysis of physical states. It is in this form that we must state the principle of causality if it is to have physical meaning. The statement that nature is governed by strict causal laws means that we can predict the future with a determinate probability and that we can push this probability as

§ 1. CAUSALITY AND PROBABILITY

close to certainty as we want by using a sufficiently elaborate analysis of the phenomena under consideration. With this formulation the principle of causality is stripped of its disguise as a principle *a priori*, in which it has been presented within many a philosophical system. If causality is stated as a limit of probability implications, it is clear that this principle can be maintained only in the sense of an empirical hypothesis. There is, logically, no need for saying that the probability of predictions can be made to approach certainty by the introduction of more and more parameters. In this form the possibility of a limit of predictability was recognized even before quantum mechanics led to the assertion of such a limit.[2]

The objection has been raised that we can know only a finite number of parameters, and that therefore we must leave open the possibility of discovering, at a later time, new parameters which lead to better predictions. Although, of course, we have no means of excluding with certainty such a possibility, we must answer that there may be strong inductive evidence against such an assumption, and that such evidence will be regarded as given if continued attempts at finding new parameters have failed. Physical laws, like the law of conservation of energy, have been based on evidence derived from repeated failures of attempts to prove the contrary. If the existence of causal laws is denied, this assertion will always be grounded only in inductive evidence. The critics of the belief in causality will not commit the mistake of their adversaries, and will not try to adduce a supposed evidence *a priori* for their contentions.

The quantum mechanical criticism of causality must therefore be considered as the logical continuation of a line of development which began with the introduction of statistical laws into physics within the kinetic theory of gases, and was continued in the empiricist analysis of the concept of causality. The specific form, however, in which this criticism finally was presented through Heisenberg's principle of indeterminacy was different from the form of the criticism so far explained.

In the preceding analysis we have assumed that it is possible to measure the independent parameters of physical occurrences as exactly as we wish; or more precisely, to measure the *simultaneous values* of these parameters as exactly as we wish. The breakdown of causality then consists in the fact that these values do not strictly determine the values of dependent entities, including the values of the same parameters at later times. Our analysis therefore contains an assumption of the measurement of simultaneous values of independent parameters. It is this assumption which Heisenberg has shown to be wrong.

The laws of classical physics are throughout *temporally directed laws*, i.e., laws stating dependences of entities at different times and which thus establish causal lines extending in the direction of time. If simultaneous values of differ-

---
[2] Cf. the author's "Die Kausalstruktur der Welt," *Ber. d. Bayer. Akad., Math. Kl.* (Munich, 1925), p. 138; and his paper, "Die Kausalbehauptung und die Möglichkeit ihrer empirischen Nachprüfung," which was written in 1923 and published in *Erkenntnis* 3 (1932), p. 32.

ent entities are regarded as dependent on one another, this dependence is always construed as derivable from temporally directed laws. Thus the correspondence of various indicators of a physical state is reduced to the influence of the same physical cause acting on the instruments. If, for instance, barometers in different rooms of a house always show the same indication, we explain this correspondence as due to the effect of the same mass of air on the instruments, i.e., as due to the effect of a common cause. It is possible, however, to assume the existence of *cross-section laws*, i.e., laws which directly connect simultaneous values of physical entities without being reducible to the effects of common causes. It is such a cross-section law which Heisenberg has stated in his relation of indeterminacy.

This cross-section law has the form of a *limitation of measurability*. It states that the simultaneous values of the independent parameters cannot be measured as exactly as we wish. We can measure only one half of all the parameters to a desired degree of exactness; the other half then must remain inexactly known. There exists a coupling of simultaneously measurable values such that greater exactness in the determination of one half of the totality involves less exactness in the determination of the other half, and vice versa. This law does not make half of the parameters functions of the others; if one half is known, the other half remains entirely unknown unless it is measured. We know, however, that this measurement is restricted to a certain exactness.

This cross-section law leads to a specific version of the criticism of causality. If the values of the independent parameters are inexactly known, we cannot expect to be able to make strict predictions of future observations. We then can establish only statistical laws for these observations. The idea that there are causal laws "behind" these statistical laws, which determine exactly the results of future observations, is then destined to remain an unverifiable statement; its verification is excluded by a physical law, the cross-section law mentioned. According to the verifiability theory of meaning, which has been generally accepted for the interpretation of physics, the statement that there are causal laws therefore must be considered as physically meaningless. It is an empty assertion which cannot be converted into relations between observational data.

There is only one way left in which a physically meaningful statement about causality can be made. If statements of causal relations between the exact values of certain entities cannot be verified, we can try to introduce them at least in the form of *conventions* or *definitions;* that is, we may try to establish arbitrarily causal relations between the strict values. This means that we can attempt to assign definite values to the unmeasured, or not exactly measured, entities in such a way that the observed results appear as the causal consequences of the values introduced by our assumption. If this were possible, the causal relations introduced could not be used for an improvement of predictions; they could be used only after observations had been made in the sense

§2. PROBABILITY DISTRIBUTIONS

of a causal construction *post hoc*. Even if we wish to follow such a procedure, however, we must answer the question of whether such a *causal supplementation of observable data by interpolation of unobserved values* can be consistently done. Although the interpolation is based on conventions, the answer to the latter question is not a matter of convention, but depends on the structure of the physical world. Heisenberg's principle of indeterminacy, therefore, leads to a revision of the statement of causality; if this statement is to be physically meaningful, it must be made as an assertion about a possible causal supplementation of the observational world.

With these considerations the plan of the following inquiry is made clear. We shall first explain Heisenberg's principle, showing its nature as a cross-section law, and discuss the reasons why it must be regarded as being well founded on empirical evidence. We then shall turn to the question of the interpolation of unobserved values by definitions. We shall show that the question stated above is to be answered negatively; that the relations of quantum mechanics are so constructed that they do not admit of a causal supplementation by interpolation. With these results the principle of causality is shown to be in no sense compatible with quantum physics; causal determinism holds neither in the form of a verifiable statement, nor in the form of a convention directing a possible interpolation of unobserved values between verifiable data.

§ 2. The Probability Distributions

Let us analyze more closely the structure of causal laws by means of an example taken from classical mechanics and then turn to the modification of this structure produced by the introduction of probability considerations.

In classical physics the physical state of a free mass particle which has no rotation, or whose rotation can be neglected, is determined if we know the *position q*, the *velocity v*, and the *mass m* of the particle. The values q and v, of course, must be corresponding values, i.e., they must be observed at the same time. Instead of the velocity v, the *momentum* $p = m \cdot v$ can be used. The future states of the mass particle, if it is not submitted to any forces, is then determined; the velocity, and with it, the momentum, will remain constant, and the position q can be calculated for every time t. If external forces intervene, we can also determine the future states of the particle if these forces are mathematically known.

If we consider the fact that p and q cannot be exactly determined, we must replace strict statements about p and q by probability statements. We then introduce *probability distributions*

$$d(q) \quad \text{and} \quad d(p) \tag{1}$$

which coordinate to every value q and to every value p a probability that this value will occur. The symbol $d(\ )$ is used here in the general meaning of *distri-*

6 PART I. GENERAL CONSIDERATIONS

bution of; the expressions $d(q)$ and $d(p)$ denote, therefore, different mathematical functions. As usual, the probability given by the function is coordinated, not to a sharp value q or p, but to a small interval dq or dp such that only the expressions

$$d(q)dq \quad \text{and} \quad d(p)dp \tag{2}$$

represent probabilities, whereas the functions (1) are probability *densities*. This can also be stated in the form that the integrals

$$\int_{q_1}^{q_2} d(q)dq \quad \text{and} \quad \int_{p_1}^{p_2} d(p)dp \tag{3}$$

represent the probabilities of finding a value of q between q_1 and q_2, or a value of p between p_1 and p_2.

Probability distributions can be determined only for sets of measurements, not for an individual measurement. When we speak of the exactness of a measurement we therefore mean, more precisely, the exactness of a type of measurement made in a certain type of physical system. In this sense we can say that every measurement ends with the determination of probability functions d. Usually d is a Gauss function, i.e., a bell-shaped curve following an exponential law (cf. figure 1); the steeper this curve, the more precise is the measurement. In classical physics we make the assumption that each of these curves can be made as steep as we want, if only we take sufficient care in the elaboration of the measurement. In quantum mechanics this assumption is discarded for the following reasons.

Whereas, in classical physics, we consider the two curves $d(q)$ and $d(p)$ as independent of each other, quantum mechanics introduces the rule that they are not. This is the cross-section law mentioned in § 1. The idea is expressed through a mathematical principle which determines both curves $d(q)$ and $d(p)$, at a given time t, as derivable from a mathematical function $\psi(q)$; the derivation is so given that a certain logical connection between the shapes of the curves $d(q)$ and $d(p)$ follows. This contraction of the two probability distributions into one function ψ is one of the basic principles of quantum mechanics. It turns out that the connection between the distributions established by the principle has such a structure that if one of the curves is very steep, the other must be rather flat. Physically speaking, this means that measurements of p and q cannot be made independently and that an arrangement which permits a precise determination of q must make any determination of p unprecise, and vice versa.

The function $\psi(q)$ has the character of a wave; it is even a complex wave, i.e., a wave determined by complex numbers ψ. Historically speaking, the introduction of this wave by L. de Broglie and Schrödinger goes back to the struggle between the wave interpretation and the corpuscle interpretation in the theory of light. The ψ-function is the last offspring of generations of wave concepts stemming from Huygens's wave theory of light; but Huygens would

§2. PROBABILITY DISTRIBUTIONS

scarcely recognize his ideas in the form which they have assumed today in Born's probability interpretation of the ψ-function. Let us put aside for the present the discussion of the physical nature of this wave; we shall be concerned with this important question in later sections of our inquiry. In the present section we shall consider the ψ-waves merely as a mathematical instrument used to determine probability distributions; i.e., we shall restrict our presentation to show the way in which the probability distributions $d(q)$ and $d(p)$ can be derived from $\psi(q)$.

The derivation which we are going to explain coordinates to a curve $\psi(q)$ at a given time the curves $d(q)$ and $d(p)$; this is the reason that t does not enter into the following equations. If, at a later time, $\psi(q)$ should have a different shape, different functions $d(q)$ and $d(p)$ would ensue. Thus, in general, we have functions $\psi(q,t)$, $d(q,t)$, and $d(p,t)$. We omit the t for the sake of convenience.

The derivation will be formulated in two rules, the first determining $d(q)$, and the second determining $d(p)$. We shall state these rules here only for the simple case of free particles. The extension to more complicated mechanical systems will be given later (§ 17). We present first the rule for the determination of $d(q)$.

Rule of the squared ψ-function: The probability of observing a value q is determined by the square of the ψ-function according to the relation

$$d(q) = |\psi(q)|^2 \qquad (4)$$

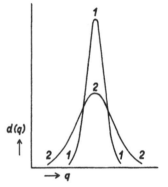

Fig. 1. Curve 1-1-1 represents a precise measurement, curve 2-2-2 a less precise measurement of q. Both curves are Gauss distributions, or *normal curves*.

The explanation of the rule for the determination of $d(p)$ requires some introductory mathematical remarks. According to Fourier a wave of any shape can be considered as the superposition of many individual waves having the form of sine curves. This is well known from sound waves, where the individual waves are called *fundamental tone* and *overtones*, or *harmonics*. In optics the individual waves are called *monochromatic waves*, and their totality is called the *spectrum*. The individual wave is characterized by its frequency ν, or its wave length λ, these two characteristics being connected by the relation $\nu \cdot \lambda = w$, where w is the velocity of the waves. In addition, every individual wave has an amplitude σ which does not depend on q, but is a constant for the whole individual wave. The general mathematical form of the Fourier expansion is explained in § 9; for the purposes of the present part it is not necessary to introduce the mathematical way of writing.

The Fourier superposition can be applied to the wave ψ, although this wave is considered by us, at present, not as a physical entity, but merely as a mathematical instrument. In case the wave ψ consists of periodic oscillations extended over a certain time, such as in the case of sound waves produced by musical instruments, the spectrum furnished by the Fourier expansion is *discrete*. Thus the individual waves of musical instruments have the wave lengths $\lambda, \frac{\lambda}{2}, \frac{\lambda}{3}, \frac{\lambda}{4}, \cdots$ where λ is the wave length of the fundamental tone and the other values represent the harmonics. In case the wave ψ consists of only one simple impact moving along the q-axis, i.e., in case the function ψ is not periodic, the Fourier expansion furnishes a continuous spectrum, i.e., the frequencies of the individual waves constitute, not a discrete, but a continuous set. As before, each of these individual waves possesses an amplitude σ, which can be written $\sigma(\lambda)$, since it depends on the wave length λ but is independent of q.

It is the amplitudes $\sigma(\lambda)$ which are connected with the momentum. We shall not try to explain here the trend of thought which led to this connection and which is associated with the names of Planck, Einstein, and L. de Broglie. Such an exposition may be postponed to a later section (§ 13). Let us suppress therefore any question of *why* this connection holds true, and let us rely, instead, upon the authority of the physicist who says that this is the case. Suffice it to say, therefore, that every wave of the length λ is coordinated to a momentum of the amount

$$p = \frac{h}{\lambda} \qquad (5)$$

where h is Planck's constant. The probability of finding a momentum p then is connected with the amplitude σ belonging to the coordinated wave λ. This is expressed in the following rule.[1]

Rule of spectral decomposition: The probability of observing a value p is determined by the square of the amplitude $\sigma(\lambda)$ occurring within the spectral decomposition of $\psi(q)$, in the form

$$d(p) = \frac{1}{h^3} |\sigma(\lambda)|^2 \qquad (6)$$

The factor $\frac{1}{h^3}$ results from the relation between p and λ expressed in (5).[2]

The two rules show clearly the connection which the ψ-function establishes between the two distributions $d(q)$ and $d(p)$, so far as it reduces these two distributions to one root. We shall later show that this kind of connection is not

---

[1] The name "principle of spectral decomposition" has been introduced by L. de Broglie, *Introduction à l'Etude de la Mécanique ondulatoire* (Paris, 1930), p. 151. In his later book, *La Mécanique ondulatoire* (Paris, 1939), p. 47, he uses also the name "principle of Born," since this principle was introduced by Born. For the rule of the squared ψ-function he uses the name "principle of interference" and in his later book the name "principle of localization".

[2] Mathematically speaking, this factor corresponds to a density function r as introduced in (22), § 9. The third power in h originates from the fact that we assume the waves to be three-dimensional.

restricted to the simple case of one mass particle, and that the same logical pattern is established by quantum mechanics for the analysis of all physical situations. For every physical situation there exists a ψ-function, and the probability distributions of the entities involved are determined by two rules of the kind described. This is one of the basic principles of quantum mechanics. We shall now construct the implications of this principle, returning once more to the simple case of the mass particle.

§ 3. The Principle of Indeterminacy

It can be shown that the derivation of the two distributions $d(q)$ and $d(p)$ from a function ψ leads immediately to the principle of indeterminacy. Let us consider a particle moving in a straight line, and let us assume that the func-

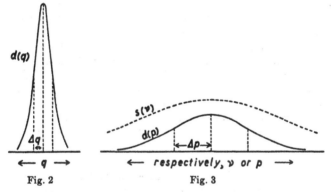

Fig. 2   Fig. 3

Fig. 2. Distribution of the position q, in the form of a Gauss curve.
Fig. 3. The dotted line indicates the direct Fourier expansion of the curve of fig. 2. The solid line is constructed through the Fourier expansion of a ψ-function from which the curve $d(q)$ of fig. 2 is derivable, and represents the distribution $d(p)$ of the momentum, coordinated to $d(q)$.

tion ψ is practically equal to zero except for a certain interval along the line. The function $|\psi(q)|^2$, i.e., the function $d(q)$, then will have the same property; let us assume that it is a Gauss curve such as is shown in figure 2. The shape of the curve means that we do not know the location of the particle exactly; with practical certainty it is within the interval where the curve is noticeably different from zero, but for a given place within this interval we know only with a determinate probability that the particle is there. Our diagram, of course, represents the situation only for a given time t; for a later time, when the particle has moved to the right, we shall have a similar curve, but it will be shifted to the right.[1]

---
[1] The curve will also gradually change its form. This, however, is irrelevant for the present discussion.

Now let us apply the principle of spectral decomposition. This decomposition, it is true, is to be applied to the complex function $\psi(q)$, and not to the real function $|\psi(q)|^2$ of our diagram. For the study of the mathematical relations in the decomposition, however, we shall first apply it to the real function of the diagram; the results can then be transferred to the complex case.

The Fourier expansion constructs the curve of figure 2 out of an infinite set of individual waves. Each of these individual waves is a pure harmonic wave of infinite length; i.e., its oscillations have a sine form and extend along the whole infinite line. Their amplitudes differ, however. The maximal amplitude will be associated with a certain mean frequency ν_0; for frequencies greater and smaller than ν_0 the amplitude will be smaller, and outside a certain range on each side of ν_0 the amplitude of the individual waves will be practically zero. Let us call the range within which the amplitudes have a noticeable size, the practical range. A bunch of harmonic waves of the kind described is also called a *wave packet*, since the superposition of all these harmonic waves results in one packet of the form given in figure 2.

Now it is one of the theorems of Fourier analysis that the practical range of a packet of harmonic waves is great when the curve of figure 2 is rather steep, but is small when this curve is flat. We can illustrate this by a diagram when we draw as abscissae the frequencies ν, and as ordinates the corresponding amplitudes $s(\nu)$ of the harmonic analysis, such as is done for the dotted line in figure 3. We choose the notation $s(\nu)$ for these amplitudes of the Fourier expansion of the real function $d(q)$, or $|\psi(q)|^2$, in order to distinguish them from the amplitudes $\sigma(\nu)$ of the Fourier expansion of the complex function $\psi(q)$. In our case, since we had assumed the curve $d(q)$ to be a Gauss curve, the curve $s(\nu)$ is also a Gauss curve, but of a much flatter shape,[2] as shown by the dotted-line curve of figure 3. This illustrates the theorem mentioned; it can be proved generally that the steeper the curve of figure 2, the flatter will be the dotted-line curve of figure 3, and vice versa. We therefore have an inverse correlation between the shape of the original curve and the shape of the curve expressing its harmonic analysis. We call this the law of *inverse correlation of harmonic analysis*.

An instructive illustration of this law is found in some problems of radio transmission. If the wave of a radio transmitter does not carry any sound, it is a pure sine wave of a sharply defined frequency. If it is *modulated*, however, i.e., if its amplitude varies according to the intensity of impressed sound waves, it no longer represents one sharp frequency, but a spectrum of frequencies varying continuously within a certain range. This range is given by the highest pitch of the sound frequencies. Consequently, a receiver with a

---

[2] The maximum of the $s(\nu)$ is at the place $\nu = 0$, and the curve is symmetrical for positive and negative frequencies ν. This results from the fact that we have assumed the curve $d(q)$ in the form of a Gauss curve. For the curve $d(p)$ there exists no such restriction, because the function $\psi(q)$ is not determined by $d(q)$, but left widely arbitrary; the maximum of $d(p)$ may therefore be situated at any value $\nu = \nu_0$, or correspondingly, at any value p.

§3. PRINCIPLE OF INDETERMINACY

sharp resonance system will pick out only a narrow domain of the transmitted waves; it will therefore drop the higher sound frequencies and reproduce transmitted music in a distorted form. On the other hand, if the resonance curve of the receiver is sufficiently flat for a high fidelity reproduction of music, the receiver will not sufficiently separate two radio stations transmitting adjacent wave lengths. The principle of inverse correlation is here expressed in the fact that it is impossible to unite high fidelity and high selectivity in the same adjustment of the receiver.

The application of these considerations to the determination of the probability distribution of the momentum of the particle involves some complications, which, however, in principle, do not change the result. As explained above, the spectral decomposition which furnishes the momenta must be applied, not to the probability curve $d(q)$ of figure 2, but to a complex ψ-curve from which this curve is derivable by the rule of the squared ψ-function. The complex amplitudes $\sigma(\nu)$ of the resulting harmonic waves then must be squared according to the rule of spectral decomposition. It is only by means of this detour through the complex domain that we arrive at the probability distribution $d(p)$ of the momentum as shown in the solid line of figure 3. Like the dotted line, this curve is also a rather flat Gauss distribution, although not quite so flat. But it can be shown that the law of inverse correlation holds as well for the two curves $d(q)$ and $d(p)$. We therefore shall speak of the *law of inverse correlation of the probability distributions of momentum and position*.

More precisely, this inverse correlation is to be understood as follows. If only the curve $d(q)$ is given, the curve $d(p)$ is not determined; it can have various forms depending on the shape of the function $\psi(q)$ from which $d(q)$ is derived. But there is a limit to the steepness of $d(p)$, represented by the solid line of figure 3. Should $d(q)$ be derived from another ψ-function than that assumed for the diagram, the resulting curve $d(p)$ can only be flatter. This general theorem can be stated without a further discussion of the question concerning the practical determination of the function $\psi(q)$ in a given physical situation. The answer to the latter question, which requires the mathematical apparatus of quantum mechanics, must be postponed to a later section (§ 20).

We said that the results obtained for the mass particle can be extended to all physical situations. Now the difference between q and p is transferred to situations in general as a difference between *kinematic* and *dynamic* parameters. We therefore shall speak generally of the *law of inverse correlation of kinematic and dynamic parameters*. It is in the form of this law of inverse correlation that we express the principle of indeterminacy. The universality of this principle follows from the fact that, whatever be the physical situation, our observational knowledge of it is summarized in a ψ-function.

The law of inverse correlation can be extended to the two parameters *time* and *energy*. This extension can be made clear as follows. A measurement of time is analogous to a determination of position. When we speak of the position

q of a particle we mean the position at a given time t; inversely we can ask for the time t at which the particle will be at a given space point q. This value t will be determinable only with a certain probability, and we can therefore introduce a probability distribution $d(t)$ analogous to $d(q)$. Similarly we can introduce a probability function $d(H)$ stating the probability that the particle will have a certain energy H. H is connected with the frequency of the harmonics by the Planck relation

$$H = h \cdot \nu \qquad (1)$$

corresponding to (5), § 2, and the probability $d(H)$ is therefore determined by the principle of spectral decomposition. Hence the two curves $d(t)$ and $d(H)$ are subject to the principle of inverse correlation in the same way as the curves $d(q)$ and $d(p)$. We shall therefore include time in the category of kinematic parameters, and energy in that of the dynamic parameters. The general law of inverse correlation of kinematic and dynamic parameters then includes the inverse correlation of time and energy.

This general law can be formulated somewhat differently when we use the concept of *standard deviation*. Let q_0 represent the mean value of q, that is, the value of the abscissa for which the curve $d(q)$ reaches its maximum; and let Δq be the standard deviation. Then the area between the curve and the axis of abscissae is divided by the ordinates $q_0 - \Delta q$ and $q_0 + \Delta q$ (drawn in figure 2) in such a way that approximately $\frac{2}{3}$ of the whole area is situated between these two ordinates. It is shown in the theory of probability that this ratio is independent of the shape of the Gauss curve. The probability of finding a value q within the interval $q_0 \pm \Delta q$ is therefore approximately $= \frac{2}{3}$. Because of these properties, the quantity Δq represents a measure of the steepness of the Gauss curve, and is therefore used as an expression characterizing the exactness of the distribution of measurements. If the standard deviation is small, the measurements are exact; if it is great, the measurements are inexact. In figure 2 and figure 3 the standard deviations Δq and Δp belonging, respectively, to the curves $d(q)$ and $d(p)$ are indicated on the axis. Now it can be shown that for the case of curves of this kind, which are derived from the same ψ-function, the relation

$$\Delta q \cdot \Delta p \geqq \frac{h}{4\pi} \qquad (2)$$

holds where h is the Planck constant. For time and energy we have the corresponding relation:

$$\Delta t \cdot \Delta H \geqq \frac{h}{4\pi} \qquad (3)$$

The inequalities (2) and (3) represent the form in which the relation of indeterminacy has been established by Heisenberg. (2) expresses the inverse cor-

§3. PRINCIPLE OF INDETERMINACY

relation of measurements of position and momentum by stating that a small standard deviation in q implies a great standard deviation in p, and vice versa. (3) states the corresponding relation between Δt and ΔH. These relations show at the same time the significance of the constant h. Since h has a very small value, the indeterminacy will be visible only in observations within the microcosmic domain; there, however, the indeterminacy cannot be neglected. The case of classical physics corresponds to the assumption that $h = 0$.

Relation (2) can be interpreted in the form: When the position of a particle is well determined, the momentum is not sharply determined, and vice versa. (3) can be interpreted in a similar way. This form makes it clear that the cross-section law of inverse correlation between kinematic and dynamic parameters states a limitation of measurability.

We now are in a position to answer the question raised above about the legitimacy of this cross-section law. If the basic principles of quantum mechanics are correct, the principle of indeterminacy must hold because it is a logical consequence of these basic principles. Furthermore, it must hold for all physical situations, because it is derivable directly from the rules of the squared ψ-function and of spectral decomposition, without reference to any special form of the ψ-function. The issue of legitimacy is reduced with this to the validity of the basic principles of quantum mechanics. Now these principles are, of course, empirical principles, and no physicist claims absolute truth for them. But what can be claimed for them is the truth of a well-established theory. Since it is a consequence of the limitation of measurability that all relations between observational data are restricted to statistical relations, we can therefore say: *With the same right with which the physicist maintains any one of his fundamental theorems, he is entitled to assert a limitation of predictability.* We may add that the same limitation follows for the determination of past data in terms of given observations, and that we therefore must also speak of a *limitation of postdictability.*

It has sometimes been said that quantum mechanics possesses a mathematical proof of the limitation of predictability. Such a statement can reasonably be meant only in the sense that there is a mathematical proof deriving the statement of the limitation from the basic principles of quantum mechanics. The principle of indeterminacy is an empirical statement; all that can be said mathematically in its favor is that it is supported by the very evidence on which the basic principles of quantum mechanics are founded. This is, however, very strong evidence.

We occasionally meet with the objection that the laws of quantum mechanics, perhaps, hold only for a certain kind of parameter; that at a later stage of science other parameters may be found for which the relation of uncertainty does not hold; and that the new parameters may enable us to make strict predictions. Logically speaking, such a possibility cannot be denied. It then might be possible, for instance, to combine a measurement of the new parameters with a measurement of the kinematic parameters in such a way that the results of measurements of the dynamic parameters could be predicted. The law of inverse correlation between kinematic and dynamic parameters then still would hold for the old parameters so long as the new ones were not used; but when the new observables were to be applied for the selection of types of physical systems, the law of inverse correlation would no longer hold within the assemblages so constructed, even for the old parameters, and therefore their values could be strictly predicted. This would mean, in other words, that we could empirically define types of physical systems for which the statistical relations controlling their parameters were not expressible in terms of ψ-functions.[3]

In such a case quantum mechanics would be considered as a statistical part of science imbedded in a universal science of causal character. Although, as we said, we cannot adduce logical reasons excluding such a further development of physics, and, although some eminent physicists believe in such a possibility, we cannot find much empirical evidence for such an assumption. If a physical principle embracing all known entities has been established, it seems plausible to assume that it holds universally, and that there is no unknown class of physical entities which do not conform to this principle. Such an inductive inference from *all known* entities to *all* entities has always been considered legitimate. The principle of describing all physical situations in terms of ψ-functions is a well-established principle, and though, certainly, quantum mechanics is still confronted by many unsolved problems and may experience important improvements, nothing indicates that the principle of the ψ-function will be abandoned. Since the relation of uncertainty and the limitation of predictability follow directly from the principle of the ψ-function, these theorems must be regarded as being as well founded in their universal claims as all other general theorems of physics.

§ 4. The Disturbance of the Object by the Observation

We now turn to considerations involving the second main issue confronting the philosophy of quantum mechanics—the issue of the interpretation of unobserved objects. This question finds a first answer in the statement that the object is disturbed by the means of observation. Heisenberg, who recognized this feature in combination with his discovery of the principle of indetermi-

---

[3] We use the term "expressible in terms of ψ-functions" in order to include both the pure case and the mixture; cf. § 23. In his book, *Mathematische Grundlagen der Quantenmechanik* (Berlin, 1932), p. 160–173, J. v. Neumann has given a proof that no "hidden parameters" can exist. But this proof is based on the assumption that for all kinds of statistical assemblages the laws of quantum mechanics, expressed in terms of ψ-functions, are valid. If the indeterminism of quantum mechanics is criticized, this assumption will be equally questioned. J. v. Neumann's proof therefore cannot exclude the case to which we refer in the text. It shows only that the assumption of hidden parameters is not compatible with a universal validity of quantum mechanics.

§ 4. DISTURBANCE OF THE OBJECT

nacy, used it as an explanation of the latter principle; he maintained that the indeterminacy of all measurements is a consequence of the disturbance by the means of observation.

This statement has aroused a wave of philosophical speculation. Some philosophers, and some physicists as well, have interpreted Heisenberg's statement as the confirmation, in terms of physics, of traditional philosophical ideas concerning the influence of the perceiving subject on its percepts. They have iterated this idea by seeing in Heisenberg's principle a statement that the subject cannot be strictly separated from the external world and that the line of demarcation between subject and object can only be arbitrarily set up; or that the subject creates the object in the act of perception; or that the object seen is only a thing of appearance, whereas the thing in itself forever escapes human knowledge; or that the things of nature must be transformed according to certain conditions before they can enter into human consciousness, etc. We cannot admit that any version of such a philosophical mysticism has a basis in quantum mechanics. Like all other parts of physics, quantum mechanics deals with nothing but relations between physical things; all its statements can be made without reference to an observer. The disturbance by the means of observation—which is certainly one of the basic facts asserted in quantum mechanics—is an entirely physical affair which does not include any reference to effects emanating from human beings as observers.

This is made clear by the following consideration. We can replace the observing person by physical devices, such as photoelectric cells, etc., which register the observations and present them as data written on a strip of paper. The act of observation then consists in reading the numbers and signs written on the paper. Since the interaction between the reading eye and the paper is a macrocosmic occurrence, the disturbance by the observation can be neglected for this process. It follows that all that can be said about the disturbance by the means of observation must be inferable from the linguistic expressions on the paper strip, and must therefore be statable in terms of physical devices and their interrelations. Quantum mechanics should not be misused for attempts to revive philosophical speculations which are not on a level with the clarity and precision of the language of physics. The solution of its philosophical problems can only be given within a scientific philosophy such as has been developed in the analysis of science and in symbolic logic.

There was a similar period in the discussion of Einstein's theory of relativity in which the relativity of time and motion was ascribed to the subjectivity of the observer. Later analysis has shown that the dependence of statements about space and time on the system of reference is in no way connected with the privacy of every person's sense data, but represents the expression of arbitrary definitions involved in every description of the physical world. We shall see that a similar solution can be given to the problems of quantum mechanics, although the situation there is even more complicated than in the case of the theory of relativity. The difference is that on the arbitrariness of definitions is superimposed, in quantum mechanics, an uncertainty in the prediction of observable results, a feature which has no analogue in the theory of relativity.

We must begin our analysis with a revision of Heisenberg's statement that the uncertainty of predictions is a consequence of the disturbance by the means of observation. We do not think that the statement is correct in this form, although it is true that there is a disturbance by the observation and that there is a logical connection between this principle and the principle of indeterminacy. This connection should rather be stated inversely, namely, in the form that the principle of indeterminacy implies the statement of a disturbance of objects by the means of observation.

To say that the indeterminacy of predictions originates from the disturbance by the instruments of observation means that whenever there is a non-negligible disturbance by observation there will always be a limitation of predictability. A consideration of classical physics shows that this is not true. There are many cases in classical physics where the influence of the instrument of measurement cannot be neglected, and where, nevertheless, exact predictions are possible. Such cases are dealt with by the establishment of a physical theory which includes a theory of the instrument of measurement. When we put a thermometer into a glass of water we know that the temperature of the water will be changed by the introduction of the thermometer; therefore we cannot interpret the reading taken from the thermometer as giving the water temperature before the measurement, but must consider this reading as an observation from which we can determine the original temperature of the water only by means of inferences. These inferences can be made when we include in them a theory of the thermometer.

Why is it not possible to apply this logical procedure to the case of quantum mechanics? Heisenberg has shown that for a precise determination of the position of a particle we need light waves of a very short wave length, that is, waves which carry rather large quanta of energy and which change the velocity of the particle by their impacts, with the consequence that this velocity cannot be measured by the same experiment. If, on the other hand, we wish to determine the velocity of a particle, we must use rather long wave lengths in order not to change the velocity to be measured; but then we shall not be able to ascertain precisely the position of the particle. If, however, the observation of a particle by illuminating it with a light ray produces an impact which throws the particle off its path, why can we not construct a theory which tells us by means of inferences starting from the result of the observation what the original velocity of the particle was? It is here that the cross-section law of Heisenberg intervenes. This principle states that whatever be the observational results, the corresponding distributions of position and momentum must be derivable from a ψ-function, and therefore must be inversely correlated. Thus, a measurement of position involves physical processes of such a kind that, relative to the observational effects of these processes, the velocity distribution is a rather flat curve. This is the reason that we cannot determine exactly the velocity of the particle in the experiment mentioned. The relation between disturbance through observation and indeterminacy must therefore be stated as follows: The disturbance by the observation is the reason that the determination of the physical entity considered is not immediately given with the measurement, but requires inferences using physical laws; since these inferences are bound to the use of a ψ-function, they are limited by the principle of indeterminacy, and therefore it is impossible to come to an exact determination. This formulation makes clear that the disturbance by the observation, in itself, does not lead to the indeterminacy of the observation. It does so only in combination with the principle of indeterminacy.[1]

In view of such objections Heisenberg's principle has sometimes been formulated as meaning: We have no exact knowledge of physical states because the observation disturbs *in an unpredictable way*. In this form the statement is correct; but then it can no longer be interpreted as substantiating the principle of indeterminacy. It *states* this principle, but does not *give a reason* for it. And we recognize that the "disturbance in an unpredictable way" is but a special case of a general cross-section law of nature stating the inverse correlation of all physical data available. The instrument of measurement disturbs, not because it is an instrument used by human observers, but because it is a physical thing like all other physical things. Instruments of measurement do not represent exceptions to physical laws; the general limitation of inferences leading to simultaneous values of parameters includes the case of inferences referring to the effects of instruments of measurement. It is in this form that we must state the principle of the disturbance by the means of observation.

This is, however, only the first step in our analysis of the disturbance by the observation. We have so far taken it for granted that we know what we mean by saying that the observation disturbs the object. In order to come to a deeper understanding of the relations involved here, we must first construct a precise formulation of this statement.

§ 5. The Determination of Unobserved Objects

When we say that the object is disturbed by the observation, or that the unobserved object is different from the observed object, we must have some knowledge of the unobserved object; otherwise our statement would be unjustifiable. Before we can enter into an analysis of the particular situation in

---

[1] A mathematical proof that the disturbance of the object by the observation does not entail the principle of indeterminacy will be given later (p. 104). The idea that it is not the disturbance in itself which leads to the indeterminacy was first expressed by the author in "Ziele und Wege der physikalischen Erkenntnis," *Handbuch der Physik*, Vol. IV (ed. by Geiger-Scheel, Berlin, 1929), p. 78. The same idea has also been expressed by E. Zilsel, *Erkenntnis* 5 (1935), p. 59. The precise formulation of the principle of uncertainty requires a qualification which will be explained in § 30.

quantum mechanics we must therefore discuss in general the problem of our knowledge of unobserved things. How do things look when we do not look at them? This is the question to which we must find an answer.

It has sometimes been said that this problem is specific for quantum mechanics, whereas for classical physics there is no such problem. This is, however, a misunderstanding of the nature of the problem. Even in classical physics we meet with the problem of the nature of unobserved things; and only after giving a correct treatment of this problem on classical grounds shall we be able to answer the corresponding question for quantum mechanics. The logical methods by which the answer is to be formulated are the same in both cases.

To begin our analysis with an example, let us assume we look at a tree, and then turn our head away. How do we know that the tree remains in its place when we do not look at it? It would not help us to answer that we can easily turn our head forward and thus "verify" that the tree did not disappear. What we thus verify is only that the tree is always there when we look at it; but this does not exclude the possibility that it always disappears when we do not look at it, if only it reappears when we turn our head toward it. We could make an assumption of the latter kind. According to this assumption the observation produces a certain change of the object in such a way that there *appears* to be no change. We have no means to prove that this assumption is false. If it is suggested that another person may observe the tree when we do not see it and thus confirm the statement that the tree does not disappear, we may restrict our assumption to cases in which no person looks at the tree, thus ascribing the power of reproducing the tree to the observation of any human being. If it is suggested that we may derive the existence of the tree from certain effects remaining observable even when we do not see the tree, such as the shadow of the tree, we may answer that we can assume a change in the laws of optics such that there is a shadow although there is no tree. The argument therefore proves only that an assumption concerning the existence, or disappearance, of the unobserved object is to be connected with an assumption concerning the laws of nature in both cases.

It would be a mistake to say that there is inductive evidence for the assumption that the tree does not disappear when we do not see it, and that this assumption is at least highly probable. There is no such inductive evidence. We cannot say: "We have so often found the unobserved tree to be unchanged that we assume this to hold always". The premise of this inductive inference is not true, since, in fact, we never have seen an unobserved tree. What we have often seen is that when we turned our head to the tree it was there; from this set of facts we can inductively infer that the tree will always be there when we look at it, but there is no inductive inference leading from these facts to statements about the unobserved tree. We therefore cannot even say that the unchanged existence of the unobserved object is at least probable.

We are inclined to discard considerations of the given kind as "nonsense",

§5. UNOBSERVED OBJECTS

because it seems so obvious that the tree is not created by the observation. Such an answer, however, does not meet the problem. The correct answer requires deeper analysis.

We must say that there is more than one true description of unobserved objects, that there is a *class of equivalent descriptions*, and that all these descriptions can be used equally well. The number of these descriptions is not limited. Thus we can easily introduce an assumption according to which the tree splits into two trees every time we do not look at it; this is permissible if only we change the optics of unobserved things in a corresponding way, such that the two trees produce only one shadow. On the other hand, we see that not all descriptions are true. Thus it is false to say that there are two unobserved trees, *and* that the laws of ordinary optics hold for them. It follows that the statements about unobserved things are to be made in a rather complicated way. Descriptions of unobserved things must be divided into *admissible* and *inadmissible* descriptions; each admissible description can be called true, and each inadmissible description must be called false. Looking for general features of unobserved things, we must not try to find *the* true description, but must consider the whole class of admissible descriptions; it is in properties of this class as a whole that the nature of unobserved things is expressed.

In the case of classical physics this class contains one description which satisfies the following two principles:

1) The laws of nature are the same whether or not the objects are observed.
2) The state of the objects is the same whether or not the objects are observed.

Let us call this descriptional system the *normal system*. It is this system which we usually consider as the "true" system. We see that this interpretation is incorrect. We may, however, make the following statement. In case a class of descriptions contains a normal system, each of the descriptions is equivalent to the normal system. If we now consider one of the unreasonable descriptions of the class, such as the statement that a tree splits into two trees whenever it is not observed, we see that these anomalies are harmless. They result from the use of a different language, whereas the description as a whole says the same as the normal system. This is the reason that we can select the normal system as the only description to be used.

The convention that the normal system be used is always tacitly assumed in the language of daily life when we speak of inductive evidence for or against changes of unobserved objects. This convention is understood when we say that our house remains in its place as long as we are absent; and the same convention is understood when we say that the girl is not in the box of the magician while he is sawing the box into two pieces, although we have seen the girl in it before. It is because of the use of this convention that ordinary statements about unobserved things are testable. The same convention is used in scientific language; it simplifies the language considerably. We must, however, realize that this choice of language has the character of a definition and that the simplicity of the normal system does not make this system "more true" than the others. We are concerned here only with what has been called a difference in *descriptive simplicity*,[1] such as we find in the case of the metrical system as compared with the yard-inch system.

In stating that a class of descriptions includes a normal system, we make a statement about the whole class. This way of stating a property of the class by means of a statement about the existence of a normal system may be illustrated by an example from differential geometry. Properties of curvature are statable in terms of systems of coordinates and their properties. Thus the surface of the sphere can be characterized by the statement that it is not possible to introduce on it a system of orthogonal straight-line coordinates which covers large areas. Only for an infinitesimal area is this possible; i.e., for small areas it is possible to introduce approximately orthogonal straight-line coordinates, and the degree of approximation increases for smaller areas. For the plane, however, such a system covering the whole plane can be introduced. It is not necessary, though, to use this "normal system" of coordinates for the plane, since any kind of curved coordinates can be used equally well; but the fact that *there is* such a normal system distinguishes the class of possible systems of coordinates holding for the plane from the corresponding class holding for a curved surface.

Similar considerations have been developed for Einstein's theory of relativity, which is the classical domain of application for the theory of classes of equivalent descriptions. Every system of reference, including systems in different states of motion, furnishes a complete description, and we have therefore in the class of systems of reference a class of equivalent descriptions. If the class of such systems includes one for which the laws of special relativity hold, we say that the considered space does not possess a "real" gravitational field. This is true, although we can introduce in such a world unreasonable systems which contain pseudogravitational fields; they are pseudogravitational because they can be "transformed away".[2]

§ 6. Waves and Corpuscles

Turning from these general considerations to quantum mechanics, we first must clarify what is to be meant by observable and by unobservable occurrences. Using the word "observable" in the strict epistemological sense, we must say that none of the quantum mechanical occurrences is observable; they are all inferred from macrocosmic data which constitute the only basis accessible to observation by human sense organs. There is, however, a class of

---
[1] Cf. the author's *Experience and Prediction* (Chicago, 1938), § 42. Descriptive simplicity is distinguished from inductive simplicity; the latter involves predictional differences.
[2] Cf. the author's *Philosophie der Raum-Zeit-Lehre* (Berlin, 1928), p. 271.

§6. WAVES AND CORPUSCLES

occurrences which are so easily inferable from macrocosmic data that they may be considered as observable in a wider sense. We mean all those occurrences which consist in coincidences, such as coincidences between electrons, or electrons and protons, etc. We shall call occurrences of this kind *phenomena*. The phenomena are connected with macrocosmic occurrences by rather short causal chains; we therefore say that they can be "directly" verified by such devices as the Geiger counter, a photographic film, a Wilson cloud chamber, etc.

We then shall consider as unobservable all those occurrences which happen between the coincidences, such as the movement of an electron, or of a light ray from its source to a collision with matter. We call this class of occurrences the *interphenomena*. Occurrences of this kind are introduced by inferential chains of a much more complicated sort; they are constructed in the form of an *interpolation* within the world of phenomena, and we can therefore consider the distinction between phenomena and interphenomena as the quantum mechanical analogue of the distinction between observed and unobserved things.

The determination of phenomena is practically unambiguous. Speaking more precisely, this means that in the inferences leading from macrocosmic data to phenomena we use only the laws of classical physics; the phenomena are therefore determinate in the same sense as the unobserved objects of classical physics. Putting aside as irrelevant for our purposes the problem of the unobserved things of classical physics, we therefore can consider the phenomena as verifiable occurrences. It is different with the interphenomena. The introduction of the interphenomena can only be given within the frame of quantum mechanical laws; it is in this connection that the principle of indeterminacy leads to some ambiguities which find their expression in the duality of waves and corpuscles.

The history of the theories of light and matter since the time of Newton and Huygens shows a continuous struggle between the interpretation by corpuscles and the interpretation by waves. Toward the end of the nineteenth century this struggle had reached a phase in which it seemed practically settled; light and other kinds of electromagnetic radiation were regarded as consisting of waves, whereas matter was assumed to consist of corpuscles. It was Planck's theory of quanta which, in its further development, conferred a serious shock to this conception. In his theory of needle radiation Einstein showed that light rays behave in many respects like particles; later L. de Broglie and Schrödinger developed ideas according to which material particles inversely are accompanied by waves. The wave nature of electrons then was demonstrated by Davisson and Germer in an experiment of a type which, a dozen or so years before, had been made by M. v. Laue with respect to X-rays, and which had been considered at that time as the definitive proof that X-rays do not consist of particles. With these results the struggle between the conceptions of waves and corpuscles seemed to be revived, and once more physics seemed to be confronted by the dilemma of two contradictory conceptions each of which seemed to be equally demonstrable. One sort of experiment seemed to require the wave interpretation, another the corpuscle interpretation; and in spite of the apparent inconsistency of the two interpretations, physicists displayed a certain skill in applying sometimes the one, sometimes the other, with the fortunate result that there was never any disagreement with facts so far as verifiable data were concerned.

An attempt to reconcile the two interpretations was made by Born who introduced the assumption that the waves do not represent fields of a kind of matter spread through space, but that they constitute only a mathematical instrument of expressing the statistical behavior of particles; in this conception the waves formulate the probabilities for observations of particles. It is this interpretation which we have used in § 2. It has turned out, however, that even this ingenious combination of the two interpretations cannot be carried through consistently. We shall describe in § 7 experiments which do not conform with the Born conception. On the other hand, the latter conception has been incorporated into quantum physics so far as it has been made the definitive form of the corpuscle interpretation. Whenever we speak of corpuscles we assume them to be controlled by *probability waves*, i.e., by laws of probability formulated in terms of waves. The duality of interpretations, therefore, is given by a wave interpretation according to which matter consists of waves; and a corpuscle interpretation, according to which matter consists of particles controlled by probability waves. As to the waves the struggle between the two interpretations, therefore, amounts to the question whether the waves have *thing-character* or *behavior-character*, i.e., whether they constitute the ultimate objects of the physical world or only express the statistical behavior of such objects, the latter being represented by atomic particles.

The decisive turn in the evaluation of this state of affairs was made by Bohr in his *principle of complementarity*. This principle states that both the wave conception and the corpuscle conception can be used, and that it is impossible ever to verify the one and to falsify the other. This indiscernability was shown to be a consequence of the principle of indeterminacy, which with this result appeared to be the key unlocking the door through which an escape from the dilemma of two equally demonstrable and contradictory conceptions was possible. The contradictions disappear, since it can be shown that they are restricted to occurrences situated inside the range of indeterminacy; they are therefore excluded from verification.

Although we should like to consider this Bohr-Heisenberg interpretation as ultimately correct, it seems to us that this interpretation has not been stated in a form which makes sufficiently clear its grounds and its implications. In the form so far presented it leaves a feeling of uneasiness to everyone who wants to consider physical theories as complete descriptions of nature; the path towards this aim seems either to be barred by rigorous rules forbidding us to

§6. WAVES AND CORPUSCLES

ask questions of a certain kind, or open only to vague pictures which have no claim to be regarded as an adequate expression of reality. It seems to us that this state of affairs is due, not so much to mistakes in the quantum mechanical part of the interpretation, as to an erroneous interpretation of the corresponding problem of classical physics which has not been seen in the full extent of its logical complications. In the following considerations we shall try to give a solution of these problems which follows the outlines of the ideas of Bohr and Heisenberg, but which seems to us to avoid the unsatisfactory parts of this conception.

We shall use for our analysis a formulation of the ideas of Bohr and Heisenberg which has been developed by Landé.[1] Landé states the duality of the two interpretations, by waves and by corpuscles, in the following form. It is incorrect to say that some experiments require the wave interpretation and others require the corpuscle interpretation; such a conception, which represents the state of affairs before the Bohr-Heisenberg theory, is not permissible, because it would make physical theory inconsistent. Instead, we must say that *all* experiments can be explained through *both* interpretations. It will never be possible to construct an experiment which is incompatible with one of the interpretations.

Combining this formulation with our theory of equivalent descriptions, and using the terminology explained above, we can state Landé's conception as follows. Given the world of phenomena, we can introduce the world of interphenomena in different ways; we then shall obtain a class of equivalent descriptions of interphenomena, each of which is equally true, and all of which belong to the same world of phenomena. In other words: In the class of equivalent descriptions of the world the interphenomena vary with the descriptions, whereas the phenomena constitute the invariants of the class. With this, the arbitrariness of descriptions is eliminated from the world of phenomena and restricted to the world of interphenomena; but there it is harmless, as we know that a similar arbitrariness of descriptions holds for the unobserved things of classical physics. Nowhere do we find an unambiguous supplementation of observations; interpolation of unobserved values can be given only by a class of equivalent descriptions.

From this result we turn to the question whether the class of equivalent descriptions of quantum mechanics contains a normal system, i.e., a description which satisfies the two principles established on page 19. Now it is obvious that the second principle will be violated by all descriptions, as there is always a disturbance of the object by the observation. We must therefore modify our definition of the normal system and restrict it to the requirement that at least the first principle be satisfied.[2] The question must therefore be asked in the

[1] A. Landé, *Principles of Quantum Mechanics* (Cambridge, England, 1937).
[2] Professor W. Pauli drew my attention to the fact that even in classical physics the second principle is mostly to be suspended for the introduction of normal systems. When we see a physical object, the fact producing the observation is the entering of light rays

form: Is there a normal system in the wider sense, i.e., a system at least satisfying the first principle?

We must not believe that the existence of a normal system can be postulated by philosophical considerations. We cannot admit that there is any *synthetic a priori* principle, i.e., a principle which is not logically empty and which physical theory is bound to satisfy. The question of whether there is a normal system can only be answered by experience. If there is such a system, the world of interphenomena is revealed to be of a rather simple structure; if there is no such system, this world turns out to be more complicated than we would perhaps like it to be. But by no means is it permissible to refuse an answer to the question as meaningless, or to evade it by drawing the attention to other sides of the problem. The general properties of the world of interphenomena, as it is constructable on the basis of quantum mechanics, are expressed in the answer we give to this question.

§ 7. Analysis of an Interference Experiment

To find an answer to the question of whether there is a normal system, let us analyze some experiments which can be considered as typical for various kinds of logical situations. First, let us consider (figure 4) the case of a diaphragm containing one slit B through which radiation of light, or electrons, or other particles of matter passes towards a screen. We then shall obtain an interference pattern on the screen. We know, however, that if we use very low intensities of the radiation, we obtain on the screen, not the whole pattern at once, but individual flashes in strictly localized areas, for instance, in C. These flashes could be verified, say, by a Geiger counter in C. If we let the experiment go on for a certain time, the distribution of the flashes, occurring one after the other, will follow the interference pattern mentioned above; it is this sum of individual flashes which a photographic film placed on the screen would show us.

The phenomena of this experiment are given by the individual flashes on the screen; besides, we have the macrocosmic objects consisting in the source of radiation, the diaphragm, and the screen. Let us ask which kind of interphenomena we can introduce here by the method of interpolation. First, we

---

into the retina of the human eye; but in this action the light ray is absorbed and thus changed by the interaction of the means of observation. The statement that there is a physical object which is undisturbed by the observation is therefore, physically speaking, obtained by an inference based on an observation which does not satisfy the second principle. Only psychologically speaking is this not true, since the inference is performed automatically by the sensory mechanism; the eye is gauged in stimulus language. This is the reason why it appears advisable in classical physics to interpret the term "observation" in such a way that the second principle can be maintained. It would be equally possible to cancel the second principle even for classical physics. This corresponds to our view that the abandonment of the second principle in quantum mechanics is rather irrelevant and that a normal system in a wider sense can be defined including the case that the second principle is violated. It is the first principle which expresses the *conditio sine qua non* of the normal system.

§7. AN INTERFERENCE EXPERIMENT

can use the corpuscle interpretation.[1] We then say that individual particles are emitted from the source of radiation, and move on straight lines such as indicated in figure 4; at the place B these particles are subject to impacts or other forms of interaction imparted to them by the particles of which the substance of the diaphragm is composed. They thus deviate from their paths. These impacts follow statistical laws in such a way that some parts of the screen are frequently hit, others less frequently. The interference pattern of the photographic film indicates, therefore, the probability distribution of the impacts given in B to the passing particles. Of course, there are also other particles

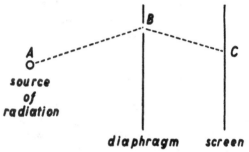

Fig. 4. Diffraction of radiation at a slit B.

leaving the source A; but when they reach the diaphragm in places other than the slit B they are absorbed or reflected and will not appear on the screen.

In this interpretation we have a certain probability

$$P(A,B) \quad (1)$$

that a particle leaving the source A will arrive at B, and a probability

$$P(A.B,C) \quad (2)$$

that a particle leaving A and passing through B will arrive at C. Both values can be statistically determined by counting on a surrounding screen all particles leaving A, then all those particles arriving on the screen (the number arrived at meaning the number of particles passing through the slit B), and then all particles arriving at C.

We see that we have here an interpretation of *interphenomena* which satisfies the first principle required for a normal system. The only deviation from classical physics consists in the fact that we have merely a probability law governing the transition from B to C; but this extension of the concept of causality holds equally for the *phenomena* of quantum mechanics. Therefore, in this interpretation both phenomena and interphenomena are governed by the same laws.

---

[1] The way of describing this experiment in corpuscle interpretation is represented in A. Landé, *Principles of Quantum Mechanics* (Cambridge, England, 1937), § 9.

Now let us use the wave interpretation. We then say that spherical waves leave A, that only a small part of these waves passes through the slit B and then spreads toward the screen. This part of the waves consists of different trains of waves, each of which has a different center; all these centers lie on points within the slit B (principle of Huygens). The superposition of these different trains furnishes the interference pattern on the screen.

So long as we consider only the results of the process obtained during a long time, for instance, in the pattern obtained on a photographic film, this explanation leads to no difficulties. It has even the advantage over the corpuscle interpretation in that it does not use statistical laws, but strictly causal laws. The numerical values of the probabilities (1) and (2) appear here as the intensities of waves which determine directly the amount of blackening in the various points of the screen. It is different as soon as we consider the individual flashes which we know can be verified on the screen. Let us assume, for instance, that the screen is replaced by a set of Geiger counters; the statistics of this system of counters then will be equivalent to the interference pattern on the film, with the addition, however, that it reveals the process as being composed of individual impacts. In face of these facts the assumption of waves leads into difficulties which were first pointed out by Einstein. So long as the wave has not yet reached the screen, it covers an extended surface, namely, a hemisphere, with its center in B; but when it reaches the screen it will produce a flash only at one point, say, at C, and will then automatically disappear at all other points. The wave is swallowed, so to speak, by the flash at C. This process of the disappearance of the wave constitutes a *causal anomaly* so far as it contradicts the laws established for observable occurrences. We see that in this description the laws of interphenomena are different from the laws of phenomena; the given description therefore does not represent a normal system.

With the intention of escaping disagreeable consequences of the wave interpretation, the suggestion has been advanced that it should be forbidden to ask questions about what becomes of the wave after a flash on the screen has been observed. We shall later discuss an interpretation in which such an interdiction of questions is carried through; with such an interpretation, however, we have abandoned the wave theory. *Within the wave interpretation* the legitimacy of such questions cannot be denied. The different reasons which have been offered for ruling out such questions do not stand a logical test. Thus it has been said that within the wave description we cannot speak of effects localized in space. This, however, is incorrect; the wave itself is conceived as a function of space, and if a certain effect is observed in one place on the screen, it is perfectly correct to ask what effects are caused by the wave in other places. It has been said, furthermore, that the flash on the screen belongs to the corpuscle interpretation, and that therefore we cannot incorporate it into the wave interpretation. This is incorrect because the flash on the screen is a verifiable phenomenon, and therefore is not included in the duality of interpretations

§7. AN INTERFERENCE EXPERIMENT

which concerns only the interphenomena. The flash belongs neither to the one nor to the other of the interpretations, but is one of the verifiable data on which both interpretations are based. Now we must demand that each interpretation of interphenomena be compatible with the given set of phenomena; and if the wave interpretation is used, it must be extended to include a statement concerning the transformation of a wave into a localized flash.

It appears understandable that in the case of such an experiment as the one considered we prefer the corpuscle interpretation, since it does not include causal anomalies, and therefore represents a normal system. But the wave interpretation is as true as the other; it is true in the same sense as is the interpretation in the example given above, according to which the tree always splits

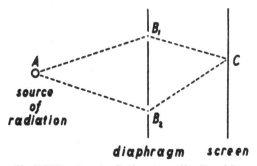

Fig. 5. Diffraction of radiation at two slits, B_1 and B_2.

into two trees at the moment it is not observed. This anomaly need not bother us because we know that it can be "transformed away" by the use of another description. In the same sense the anomaly of the wave description should not bother us because we know that it can be transformed away by the use of the corpuscle description. But if the wave interpretation is used, it includes the consequence of the disappearance of the whole wave with the appearance of the flash in one point, however distant this point may be from other points of the wave. We must have the courage to face this consequence which is necessarily combined with this interpretation of interphenomena.

Let us now turn to a second experiment, which we indicate in figure 5. We use the same arrangement as before, with the difference that the diaphragm has two slits, at B_1 and B_2. We know that in this case we shall also obtain an interference pattern on the screen, which is, however, different from the pattern obtained in the first experiment. Let us consider this experiment in different interpretations.

We first use the corpuscle interpretation. Assuming, as before, low intensities of the radiation, we know that there will be individual flashes on the screen. We can explain this by the assumption that individual particles leave the source intermittently; sometimes a particle passes through the slit B_1, sometimes through the slit B_2, sometimes it is absorbed by the diaphragm—all this being determined by the direction in which the particle leaves A. If there occurs a flash in C, we shall say that the particle has passed either through B_1 or B_2.

The probability that a particle reaches C can be stated by the rule of elimination established in the calculus of probability:

$$P(A,C) = P(A,B_1) \cdot P(A.B_1,C) + P(A,B_2) \cdot P(A.B_2,C) \tag{3}$$

The meaning of these terms follows from the explanation given with respect to (1) and (2). We should be inclined to assume that the numerical values of the probabilities on the right hand side of (3) are the same as obtained in experiments of the first type; it turns out, however, that this assumption is wrong.

This can be proved in the following way. We first close the slit at B_2, and let the process of radiation go on for a certain time; then we close the slit at B_1, and let the process go on for an equal time. Using a film as a screen we then obtain a superposition of both interference patterns. The question arises: Is this pattern of interference the same as that which will result when both slits are opened simultaneously? If it is, we can assume that the values of the occurring probabilities are the same as in the first experiment. If it is not, the probabilities $P(A.B_1,C)$ and $P(A.B_2,C)$ must have changed.

It is well known that the experiment decides in favor of the second alternative. We therefore must assume that the probability with which a particle passing through B_1 reaches C depends on whether the slit B_2 is open.[2] This is a causal anomaly; it states that there is an effect originating in B_2 and being spread to B_1 such that it influences the impacts given in B_1 to passing particles. We see that it is in this case the corpuscle interpretation which leads to causal anomalies.[3]

It would be erroneous to say that because of these anomalies the corpuscle interpretation is *false*. Some physicists who have uttered opinions of this kind have based their judgments on the fact that we have no means of knowing through which of the two slits, B_1 or B_2, the particle has passed after the flash in C has been observed. The latter statement, of course, is true, since an observation in B_1, or in B_2, would disturb the experiment. We are not even able to determine the probability $P(A.C,B_1)$ that the particle, observed in the flash C,

---

[2] It can be easily seen that this consideration is a special case of the consideration given in § 22, according to which an intermediate measurement of an entity v influences the probability leading from u to w, even if the result of the measurement is not included in the statement of the latter probability. Closing the slit B_2 is equivalent to making a measurement of position in B_1.

[3] Further anomalies arise if we bring the screen closer to the diaphragm. If we choose the point C on the screen always in such a way that the direction B_1C remains the same, we find that the probability $P(A.B_1,C)$ cannot be considered as remaining constant; i.e., this probability is not a function depending only on the direction in which the particle leaves B_1. This follows from considerations using waves.

§7. AN INTERFERENCE EXPERIMENT

has passed through B_1. This inverse probability could be determined by means of the rule of Bayes[4] in the form

$$P(A.C,B_1) = \frac{P(A,B_1) \cdot P(A.B_1,C)}{P(A,C)} \quad (4)$$

if we knew the value of the forward probability $P(A.B_1,C)$; but since this probability is different from the value $P(A.B,C)$ obtained with the use of only one slit, it cannot be determined. Every such determination would require observations of the particle in B_1, and therefore would lead to a disturbance of the experiment. Applying the verificability theory of meaning, even in the modified form of probability meaning,[5] we therefore must say that the sentence "the particle has passed through B_1" is meaningless if we consider it as a *statement about physical facts*. It is permissible, however, to use this sentence if we consider it as a *definition*. In order to make the corpuscular description complete we coordinate by definition to every flash in C a path of a particle, passing either through B_1 or B_2; the choice of these paths is arbitrary. Even if we follow the requirement (which in itself has only the character of a definition) that the probability $P(A,B_1)$ and $P(A,B_2)$ be the same as obtained in experiments with one slit, the choice of the paths will remain arbitrary within wide limits.

The situation presented reminds us of a similar situation in the problem of simultaneity. If a light signal leaving a point Q at the time t_1 is reflected at a point R, and then returns to Q at the time t_3, we may coordinate to the time of its arrival in R every numerical value between t_1 and t_3. With this choice we make a definition of simultaneity at the points Q and R. The sentence: "one of the events in Q between t_1 and t_3 is simultaneous with the arrival of the light signal in R" is meaningless if it is considered as an empirical statement, since it is not verifiable; but it is meaningful if we introduce it as a definition. In the same sense we can use the sentence: "the particle passed through B_1" as a definition. In both cases such a definition is necessary in order to make our description complete.

It follows that the corpuscular interpretation can be carried through consistently, and that there is nothing incorrect in it. Its only disadvantage is that it leads to causal anomalies as explained. The influence of the slit B_2 on the occurrences in B_1 is of such a kind as to violate the principle of *action by contact*. There is no spreading of the effect from B_2 to B_1; this can be seen from the fact that changes in the material of the diaphragm between B_1 and B_2, or changes of its shape such as would result from corrugating the diaphragm, would not influence the effect.

If we want to construct for the experiment described an interpretation free

---
[4] Cf. any textbook on probability, or the author's *Wahrscheinlichkeitslehre* (Leiden, 1935), formula (6), § 21.
[5] Cf. the author's *Experience and Prediction* (Chicago, 1938), § 7.

from anomalies, we must use a wave interpretation. It would not be sufficient, however, to use an interpretation in which the wave spreads through all the open space; this would lead to the same anomalies as explained for the wave interpretation of the first experiment. We must use a wave interpretation according to which the wave is limited to two narrow canals such as indicated in figure 6. It can be shown that any changes made at a point inside the "two-canal element" would produce a change in the probability $P(A,C)$ that a flash in C occurs, whereas changes outside do not influence this probability. We could, for instance, fill the space outside the strips included in the dotted lines with absorbing matter, without changing $P(A,C)$.[6] This can be proved through

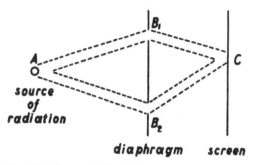

Fig. 6. The "two-canal element" representing the radiation.

the principle of superposition which we used above; the interference pattern on the screen can be considered as the superposition of the blackenings, produced one after the other, by two-canal elements having their top ends in different points C on the screen. We therefore obtain a normal system for the considered experiment if we use a description in which the interphenomena are regarded as two-canal waves spreading from A to various points C on the screen, and following each other intermittently.

We may add here the remark that for the case of only one slit, i.e., for the experiment of figure 4, we have also a wave interpretation which is free from causal anomalies; we then speak of a one-canal wave, such as Einstein originally assumed in his needle radiation. We therefore have here two normal systems. But the difference between these two descriptions is not very great, and we therefore usually speak only of the corpuscle interpretation. Every statement about corpuscles can therefore be replaced by a statement about needle radiation.

[6] Strictly speaking, this is only approximately true. The degree of approximation increases when the canals chosen are wider in their middle parts, while the widths at B_1, B_2, and C remain unchanged. The mathematical theorem to which we refer here consists in the statement that it is possible to introduce two sets of waves starting from B_1 and B_2 interfering on the screen in such a way that only in a small area at C an intensity is left. Our "two-canal element" may be considered as a simplified illustration of these two sets of waves.

§7. AN INTERFERENCE EXPERIMENT

It has sometimes been said that the differences of wave and corpuscle interpretation are combined with an alternative concerning the use of space-time concepts and causal concepts. The wave interpretation, according to this conception, satisfies the principle of causality so far as the waves are controlled by a differential equation, the Schrödinger equation (cf. § 13), but does not allow us to give a spatio-temporal description of physical objects. The corpuscle interpretation, on the other hand, is said to satisfy the requirements of a spatio-temporal description but to violate the principle of causality. Apart from the latter statement we cannot consider this conception as correct. It is not true that the wave description satisfies throughout the conditions of causality. It does so only so far as the wave field, conceived as a physical reality, spreads through space in a form expressible by a differential equation; in this respect it represents an action by contact, at least when free particles are concerned.[7] But, as we have shown, there are other points in which the latter principle is violated. Thus, the disappearance of the wave after the occurrence of a flash on the screen is a process not following the Schrödinger equation, and therefore not conforming to the principle of action by contact. In addition, we must say that a wave interpretation which does not satisfy the conditions of a space-time description *eo ipso* cannot satisfy the postulates of a normal causality either. Spatio-temporal order is closely connected with causal order, as has been brought to light in the analysis of the theory of relativity.[8] If the wave cannot be imagined as imbedded in a space-time manifold in which *every* part of the wave process satisfies the principle of action by contact, it cannot be said to conform to the requirements of a normal causality.

Our exposition has shown that the problems involved in quantum mechanics cannot be reduced to the alternative space-time versus causality. The space-time order to be assumed is always the ordinary one, and can be ascertained in many cases by the usual macrocosmic methods; thus the distance between the two slits B_1 and B_2 may be large enough to be measured by macrocosmic appliances. In both interpretations it is violations of the requirements of normal causality which are concerned. The kind of these violations only varies with the interpretation.

Such violations will occur also within a third interpretation which we must mention now and which is given by a combination of both wave and corpuscle interpretation. According to this interpretation we have a wave field spreading through the whole space and being diffracted at the slits B_1 and B_2 in the usual way; in addition to this field we have corpuscles whose movement is controlled

---

[7] For systems composed of several particles the waves spread, not in a three-dimensional space, but in the n-dimensional configuration space. But even if we were to consider the configuration space as "real" space, the waves would not satisfy the requirements of normal causality. In such a space the disappearance of the wave after the flash would lead to the same difficulties as in ordinary space.

[8] Cf. the author's *Philosophie der Raum-Zeit-Lehre* (Berlin, 1928), §§ 27, 42.

by the field in such a way that the intensity of the field determines the probability of finding a corpuscle. Using a term introduced by L. de Broglie, we speak here of *pilot waves*. This interpretation has its anomalies in the existence of a field which follows laws different from those holding for other kinds of waves; in particular, this wave field possesses no energy, since the energy is supposed to be concentrated in the particles. Furthermore, the influence of the field on the particles is governed by unusual laws. If the particles are assumed to move in straight lines, these laws would violate the principle of action by contact, since, then, the probability that a particle passing through B_1 will turn in the direction of C would depend, not on the intensity of the field near B_1, but on its intensity at C.[9] Assuming that the particles move in oscillating lines we encounter other anomalies.[10]

This description is therefore not a normal system, but it has some advantages which make its application advisable in many cases. It is not necessary to assume that the pilot waves appear intermittently with the particles, and that these waves disappear with the flash in C. The waves may be assumed to go on continuously; their existence, however, will be verifiable only at the moments when there are particles traveling through them.

§ 8. Exhaustive and Restrictive Interpretations

The above analysis shows that neither the corpuscle interpretation nor the wave interpretation can be carried through without causal anomalies. Using the particle interpretation we can explain some experiments in such a way that the laws of phenomena and interphenomena are the same; but then we encounter anomalies in other cases. Using the wave interpretation we can explain these other cases in such a way that the laws of phenomena and interphenomena are the same; but then anomalies appear in the explanation of experiments of the first kind. Finally, a combination of the two into an interpretation of pilot waves shows other anomalies.

The question arises whether there is another interpretation, perhaps unknown to us, which is free from causal anomalies. The preceding investigations cannot be considered as a proof that there is no such interpretation. Such a proof cannot be given by trying out one interpretation after another; we then are never sure whether a better interpretation which escaped our attention remains. The proof must be based on a general theory of the relations between quantum mechanical entities. We shall give this proof in § 26; the conception of causality assumed for it, which takes account of possible modifications of this notion, will be explained at the end of § 24. Our results can be formulated as follows: It is impossible to give a definition of interphenomena in such a

---

[9] This follows from considerations similar to those indicated in fn. 3, p. 28.
[10] The anomalies of this interpretation are very clearly presented in L. de Broglie's book, *Introduction à l'Etude de la Mécanique ondulatoire* (Paris, 1930), chap. 9.

§8. KINDS OF INTERPRETATIONS

way that the postulates of causality are satisfied. *The class of descriptions of interphenomena contains no normal system.* This can be proved to be a consequence of the basic principles of quantum mechanics. We shall call this result the *principle of anomaly*.

In view of this negative result two different conceptions can be carried through. The first calls for a *duality of interpretations*. Among the class of equivalent descriptions we have two, the corpuscle interpretation and the wave interpretation, which are more expedient than the others; since we have no normal system, we can use, instead, either of these two interpretations as a *minimum system*, i.e., a system for which the deviations from a normal system constitute a minimum. In this conception causal anomalies cannot be avoided; but they can at least be reduced to a minimum.

The second conception represents a more radical remedy. Since no normal description of interphenomena exists, it has been suggested we should renounce any description of interphenomena; we should restrict quantum mechanics to statements about phenomena—then no difficulties of causality will arise. The impossibility of a normal system is construed, in this conception, as a reason for abandoning all descriptions of interphenomena. We shall call conceptions of this kind *restrictive interpretations* of quantum mechanics, since they restrict the assertions of quantum mechanics to statements about phenomena. The rule expressing this restriction can assume various forms, and we shall therefore have several restrictive interpretations. Interpretations which do not use restrictions, like the corpuscle and the wave interpretation, will be called *exhaustive interpretations*, since they include a complete description of interphenomena.

The adherents of restrictive interpretations have maintained that a description of interphenomena is unnecessary; for the purpose of observational predictions, they say, it is sufficient to have an interpretation which refers only to phenomena. The latter statement is true; but it cannot be considered as proof that exhaustive descriptions should be abandoned. We should clearly keep in mind that neither of the two conceptions can be proved to be true. These conceptions represent volitional decisions concerning the form of physics; either of them is as justifiable as the other.

Speaking in terms of the class of equivalent descriptions, the situation can be characterized as follows. The system of phenomena is the same for each description of this class; it is therefore the *invariant* of this class. Now the class is dependent on its invariant; so far, any restrictive interpretation determines the whole class of exhaustive descriptions. The latter descriptions, however, reveal a feature which we would not know if we knew only a restrictive description: this is the fact that no interpretation free from causal anomalies can be given. Since this is a property of the class of exhaustive descriptions, it represents an inherent property of every restrictive interpretation. This property is expressed in the restrictive interpretations through the fact that they rule out certain statements; but the reason for this rule can only be formulated in terms of a statement about the properties of the class of exhaustive descriptions.

We therefore shall turn now to a further analysis of the class of exhaustive interpretations, while we postpone the discussion of restrictive interpretations. Within the first class, we said, the two interpretations in terms of corpuscles and waves hold a special position so far as they represent minimum systems. To this we now must add a second statement which secures a unique position to these two interpretations, and which at the same time attenuates the consequences resulting from the absence of a normal system.

Although we have no exhaustive description free from anomalies holding for *all* interphenomena, we can construct such a description for *every* interphenomenon by using either the wave or the corpuscle interpretation. It is this fact which we express in speaking of the *duality* of wave and corpuscle interpretation. We mean by this that *for a given experiment* at least one of the two will be a normal description and will thus define interphenomena in such a way that they follow the same laws as the phenomena; it is only in other experiments that the interpretation so chosen will lead to causal anomalies. Let us call this statement the *principle of eliminability of causal anomalies*. The difference between *all* and *every*, which we used to formulate this principle, is well known to symbolic logic. Using this grammatical distinction in another form we may also say: It is false to say that *all* interphenomena follow the laws holding for phenomena; but it is correct to say that *every* interphenomenon does so. *We do not have one normal system for all interphenomena, but we do have a normal system for every interphenomenon.*

As before, an analogy from differential geometry may illustrate these formulations. When we use a system of orthogonal coordinates on the sphere, such as that given in the circles of longitude and latitude (such a system is possible because it does not consist throughout of straightest lines, the circles of latitude not being greatest circles), this system has singularities at the North Pole and the South Pole; i.e., these points do not have a definite longitude. These singularities, however, are due only to the system of coordinates; the poles themselves are not distinguished geometrically from any other point of the sphere. The singularities can therefore be "transformed away" by the introduction of another system of coordinates; thus, the sailor will use, near the poles, a notation of points which determines positions relative to a chosen initial point and two chosen directions rectangular to each other. These coordinates could even be produced and used to cover the whole sphere, at least if one set of lines is not assumed to consist of straightest lines; but then singularities will appear in other points of the system. We may call a system of orthogonal coordinates without singularities a normal system. Then we may say that we can introduce a normal system for *every* extended area on the sphere, but we cannot introduce one normal system for *all* areas, i.e., for the

§8. KINDS OF INTERPRETATIONS

whole sphere. We thus express a statement about the sphere in terms of a statement concerning the class of possible systems of coordinates.

The difference between this case and the case considered above, which concerns orthogonal straight-line coordinates, is as follows. A system of coordinates being both orthogonal and straight-lined is possible only for infinitesimal areas; for extended areas of some size it cannot even be carried through approximatively. If we renounce the requirement of straight lines we can construct a system which covers great areas of the sphere, and which is strictly orthogonal; but such a system will lead to singularities in two points. The advantage of this case over the first is that, with this definition of the normal system, we obtain a normal system for extended areas, not for infinitesimal areas only.

Returning to quantum mechanics, we must say that the situation there corresponds to the second case. The causal anomalies can be transformed away strictly for "extended areas", i.e., for a whole experiment, by a suitable description. They will reappear only for other experiments or for questions in which experiments of different kinds are compared; for the answer to such questions we then can introduce a new description such that once more the anomalies disappear. The reason that this is always possible is given in the relation of indeterminacy. If we could observe a particle passing through the slit B_1 in the experiment of figure 5, we could not introduce a wave description, and therefore would have no normal description of the experiment, i.e., no description free from anomalies. On the other hand, if we could prove that a wave arrived simultaneously at different points C of the screen in the experiment of figure 4, we could not introduce a corpuscle description, and therefore would have no normal description of this experiment. We see that *the principle of eliminability of anomalies is made possible through the principle of indeterminacy*, since the latter principle makes it impossible ever to construct a crucial experiment between wave interpretation and corpuscle interpretation.[1]

The ultimate root of the duality of wave interpretation and corpuscle interpretation is therefore given in the principle of indeterminacy; but this principle also points the way out of the dilemma of causal anomalies, a result stated by us in the principle of eliminability. We spoke above of the skill displayed by physicists in applying sometimes the wave interpretation and sometimes the corpuscle interpretation; we now see that we can give a justification of this change of interpretations which proves that the switching over to a normal interpretation is a legitimate means of physical analysis. When the physicist, in face of a particular experiment, introduces a suitable description which eliminates causal anomalies within the frame of his question, he may be compared

---

[1] It may be questioned whether it is actually possible in all cases to eliminate causal anomalies by a suitable description. What can be shown is that the principle of eliminability holds, at least for single particles or for swarms of particles which do not interact with each other such as electron swarms or light rays. Difficulties arise for complicated structures composed of several particles. Cf. § 27.

to the sailor who, at the North Pole, discontinues determining his position in terms of longitude, and prefers to use another system of coordinates free from singularities. Such a procedure is permissible because nature has not determined one normal system for all interphenomena, but only a separate normal system for each interphenomenon.

Let us consider some examples. Using the wave interpretation, we arrive at the question why the whole wave disappears after a flash has been observed on one point of the screen. We eliminate the causal anomaly presented in this description of the interphenomenon by introducing the particle interpretation. The wave then is transformed into a probability, and, instead of the disappearance of a wave, we have the simple statement that although the probability $P(A, C_2)$ of finding a flash on the screen in a point C_2 has a certain positive value, the probability $P(A.C_1, C_2)$ of finding a flash in C_2 after a flash has been observed in C_1, is zero. Instead of the contraction of a wave into a point, we have here the trivial logical fact that probabilities are relative. It was by considerations of this kind that Born was led to the introduction of the statistical interpretation of the waves which originally had been conceived by Schrödinger as waves of electrical density.

Another example where the normal description is given by the particle interpretation is represented by the following consideration. The probability that a particle leaving the source A will pass through either of the slits is the sum of the probabilities that a particle will pass through one of the slits; we may indicate this by the symbolic expression:

$$P(A, B_1 \vee B_2) = P(A, B_1) + P(A, B_2) \qquad (1)$$

(The logistic sign "$\vee$" means "or".) This relation follows from the rules of probability because the particles can go through only one of the holes at a time. It can be tested by observations, as follows: To ascertain the value of the left hand side we count all flashes occurring on the screen when both slits are open; to determine the two values of the right hand side we count all flashes on the screen occurring, respectively, when one of the slits is closed. Since statistics so compiled show that (1) holds, this relation must hold also for the wave interpretation. Here, however, the explanation leads to causal anomalies. We then must assume that whenever a wave leaves A in the direction of B_1 there is another wave leaving simultaneously in the direction of B_2, but that the wave going toward the open slit B_1 will sometimes disappear due to an influence of the closed slit B_2 (namely, in all those cases when, in the corpuscle interpretation, a particle is emitted only in the direction of B_2), and is thus controlled by an influence which represents an action at a distance. The physicist who wishes to explain the well-confirmed relation (1) will therefore prefer the corpuscle interpretation by the use of which he can derive the relation without the assumption of anomalies.

On the other hand, there are questions which only by the use of the wave

§8. KINDS OF INTERPRETATIONS

interpretation can be answered without reference to anomalies. We saw that the corpuscle interpretation of the experiment indicated in figure 5, § 7, involves an action at a distance between the two slits B_1 and B_2. In the wave interpretation this action at a distance is eliminated and replaced by a statement about a phase relation between the waves arriving in B_1 and B_2, which is due to their common origin from the source A. In answering questions of this kind the physicist will therefore prefer the wave interpretation.

Our examples show that it is even preferable to speak, not of the normal system for every interphenomenon, but, of the normal system *for every question* concerning interphenomena. It is the question which determines the normal system, and, relative to the same experimental arrangement, different questions can be asked which require different normal systems. Thus the question concerning the probability relation (1) is asked with respect to an experimental arrangement which, for other questions, necessitates a wave interpretation. When we say "for every question", we mean, of course, that the question is sufficiently limited, and not constructed as an "and"-combination of different questions. With this qualification we can formulate the principle of eliminability as stating: We have no normal system for all questions concerning interphenomena, but we do have a normal system for every such question.

The switching over from one interpretation to another is justifiable, we said, as a means of eliminating causal anomalies. This is, however, its only justification, and it would be incorrect to adduce reasons of another kind. We sometimes read that questions like "what becomes of the wave after a flash on the screen has been observed" or "why does a particle going through the slit B_1 move differently according as the slit B_2 is closed or open" must be forbidden because they are not *adequate* to the respective interpretation. But this word "adequate" means only that the answer to such questions leads to causal anomalies. The occurrence of such anomalies does not make the question, or the answer, unreasonable. If we have decided to use one of the exhaustive interpretations, such questions are not meaningless. We then must become accustomed to the fact that for a given interpretation there are always questions which can only be answered by the assumption of causal anomalies. If we prefer to use in the answer to such questions an interpretation which is free from anomalies, we have good reasons to do so; but we should not believe that the answer so constructed is *the only meaningful* answer, or *the only admissible* answer, or *the only true* answer. All the merits of such an interpretation consist in the fact that it is free from causal anomalies for the interphenomenon considered; but neither does this fact make it more true than others, nor is it a necessary condition of meaning within an exhaustive interpretation. Judgments of this kind are based on a confusion of exhaustive and restrictive interpretations. Only for the latter will the mentioned questions be meaningless; but for such an interpretation the complete description of the experiment which is free from anomalies is meaningless as well. Within an exhaustive description, however, we shall be entitled to state causal anomalies as well as cases of normal causality.

On the other hand, the opposite mistake has been frequently made, i.e., the mistake of saying that we have no true description at all of the interphenomena; and that the duality of interpretations proves that we can construct only pictures of the interphenomena, correct in some features and incorrect in others. Whereas the previously criticized attitude appears dogmatic, the latter attitude must be called too modest. All admissible descriptions are equally true in all their details, including the anomalies. If somebody wishes to call descriptions of interphenomena *pictures* he may do so in order to stress the possibility of choice; but he must not forget that then it is also a use of pictures to say that the tree on the street remains in its place when nobody looks at it. We saw that in this case also we have a choice of descriptions, and that the question of the unobserved tree can be unambiguously answered by inductive evidence only after the postulate of unchanged laws of nature has been introduced. It is only the combination of this postulate with a description of the unobserved object which can be empirically verified. For the interphenomena of quantum mechanics the situation is the same; an unambiguous statement concerning interphenomena can be made only after the postulate of unchanged laws of nature has been introduced, and the combination of a description of interphenomena with this postulate can be empirically verified. It will determine one interpretation for every experiment, but not one for all.

If once the problem of the description of interphenomena has been formulated in this way, it is clear that even in the macrocosm there is no logical need for the existence of a normal system. Imagine that whenever we turn our eyes away from a tree and observe only its shadow we see two equally shaped shadows, whereas when we see both the tree and its shadow there is only one tree and one shadow. We then have the choice of saying either that there are always two trees when we do not look at a tree, or that there is only one unobserved tree for which, however, the known laws of optics do not hold. Therefore one of the two principles defining the normal system in the classical sense must be abandoned in such a case. Imagine, furthermore, observations showing that all changes happening to one of the shadows take place in the other shadow as well; for instance, if we feel a blast of wind and see a certain branch of one of the tree shadows being moved, we see an equal movement in the corresponding branch of the other tree shadow; if the shadow of a bird appears on one of the tree shadows, an equal shadow of a bird appears on the other tree shadow, etc. If, in this case, we want to use a description in which the laws of optics are unchanged, we must assume a duplication of occurrences, which would represent a preëstablished harmony, or an action at a distance, which produces duplicates of occurrences in different places. Since this assumption signifies a causal anomaly, we have, in the case considered, no normal description in the sense of a description satisfying the first of our principles.

§8. KINDS OF INTERPRETATIONS 39

A macrocosmic analogy resembling more closely the experiment of figure 5, § 7, can be constructed as follows. Imagine in A a machine gun which is turned irregularly by a machine so that it shoots bullets in all directions in an irregular sequence. Let the diaphragm have a certain thickness so that on the walls of the short slit canals the bullets can be reflected. We then shall observe an irregular distribution of bullets on the screen which may consist of lead so that the bullets are caught in it. Imagine, furthermore, that the bullets move so fast that we cannot see through which hole they pass, and that we have no other means to verify this. Let us now assume the following observations to be made. If both slits are open, the number of bullets hitting the screen is twice as large as in the case when only one slit is open. On certain spots of the screen, however, we find no hits whenever both slits are open, whereas we do find hits on these spots when only one of the slits is open. In such a case we have the same choice of interpretations as in the case of the radiation experiment described above. We may assume that the bullets on their path through the air remain individual particles, but that there is an action at a distance between the two slits; or that the interphenomena consist in waves spreading through both holes and uniting later to form the bullets found in the lead of the screen.

Such analogies may make it clear that there is nothing unimaginable in the state of affairs ascertained by quantum mechanics, and that it is possible to construct macrocosmic models of it. The physics of these models, of course, will be different from that of our actual macrocosm. But should our macrocosm follow a similar pattern, we should after some time get accustomed to it; we should consider it as a matter of course that we could not give one normal description for all interphenomena, and should learn to use, for the purpose of answering a certain question, the description which at least for that question does not involve anomalies. Fortunately our daily world does not show this kind of structure. It is different with the atomic world; quantum mechanics has shown that its structure is of the kind depicted in these analogies.

This means that in the world of atomic dimensions the postulate of unchanged laws of nature cannot be carried through for the totality of interphenomena, and therefore does not determine one interpretation as the normal interpretation of all interphenomena, although it determines one normal interpretation for every interphenomenon. This result must be considered as the most general statement which physics, in its present status, can make about the structure of the physical world. It may seem strange that the physical world cannot be caught in the network of one normal description; that the idea of uniformity of nature, so often claimed to be the ultimate result of science, cannot be extended to include the interphenomena of the world of quanta. We are dealing here, however, with a question which must be answered, not by wishful thinking, but by experimental inquiry. Since physics has come to the result as stated we must now take it seriously, and not palliate it by calling it a breakdown of human capacities for inventing pictures.

40   PART I. GENERAL CONSIDERATIONS

In order to be complete in our statement of general properties of the physical world, we must add to this negative result the positive statement formulated in the principle of eliminability. Nature allows us to construct, at least partly, the world of interphenomena in agreement with the laws of phenomena. This fact has consequences of great bearing. One is that we can answer all questions by constructing suitable interphenomena which follow normal laws. Another is that the anomalies incurred with other descriptions can never be used to produce anomalous effects in the world of phenomena. Thus we cannot use the action at a distance existing between the two slits B_1 and B_2 in the corpuscle description in order to send signals from one slit to the other. No such anomalous effect is possible because otherwise it would also occur in the normal description of the experiment; there, however, it is excluded through the normal behavior of the interphenomena. The causal anomalies which we encounter in anomalous descriptions may therefore be considered as pseudo anomalies; they are due to the form of the chosen description and can be eliminated. We see here the far-reaching significance of the principle of indeterminacy: It reveals the discrepancy between the laws of phenomena and the laws of interphenomena as not being of a malignant nature. The causal anomalies have a ghostlike existence; they can always be banished from the part of the world in which we happen to be interested, although they cannot be banished from the world as a whole.

It is this specious character of the causal anomalies which suggests the use of restrictive interpretations. Every exhaustive interpretation states too much so far as it speaks of causal anomalies which have no bearing upon the world of observable phenomena. It may therefore seem advisable to renounce exhaustiveness, and to prefer a restrictive interpretation which is free from statements involving such causal anomalies. We are thus led back to the problem of restrictive interpretations to which we must now turn for closer consideration.

A restrictive interpretation has been introduced by Bohr and Heisenberg. The *rule of restriction* states that only statements about measured entities, i.e., about phenomena, are admissible; statements about unmeasured entities, or interphenomena, are called meaningless. This has the immediate consequence that statements about the simultaneous values of complementary entities cannot be made. The interpretation so introduced is neither a corpuscle nor a wave interpretation; since it leaves the status of interphenomena widely indeterminate, we cannot say whether these interphenomena consist of particles or of waves. It resembles a corpuscle interpretation when the measured entity is the position, since then one rather sharply defined space point is attributed to the entity. But since a statement about the momentum is left open, we do not know whether the entity so determined is a particle. If we consider the whole spectrum of possible momenta as simultaneously realized, the entity might as well be a wave packet. On the other hand, if a measurement of mo-

§8. KINDS OF INTERPRETATIONS 41

mentum is made, we can consider this value either as the momentum of a particle or as the frequency of a wave; the restrictive interpretation leaves this question open.

Let us consider an example in order to show the rule of restriction at work. The interference experiment of figure 5, § 7, is an arrangement which allows us to measure a frequency, and therefore the momentum of a particle; so, all statements about the position of the particle are ruled out. This means that not only a statement of the kind "the particle went through slit B_1," is inadmissible, but that even a statement of the form "the particle went either through slit B_1 or through slit B_2" is forbidden. It is clear that this rule works. We then cannot say that *if* the particle went through slit B_1 it must have been influenced by the existence of slit B_2; the clause with "if" belongs to the forbidden domain. The causal anomaly has therefore disappeared from the domain of admissible statements. The rule of restriction, like a surgical operation, cuts off all unhealthy parts of quantum mechanical language. Unfortunately, like all such operations, it also cuts off some sound parts. Thus it is hard to abandon a statement like the one concerning the particle's going through one slit or the other. All that can be said against this statement is that it leads to undesired consequences.

It should be realized that the elimination of causal anomalies is the only justification of the restrictive interpretation of Bohr and Heisenberg. If it were possible to construct an exhaustive interpretation free from causal anomalies, no one would question the legitimacy of the definitions used in such an interpretation, even if the relation of indeterminacy should hold and make it impossible to replace these definitions by verifiable statements. Thus, if the experiment should show that the interference pattern on the screen in figure 5, § 7, were equal to the superposition of the two interference patterns resulting when first one slit is open and then the other, no one would doubt that the particles went either through one slit or the other, although, because of the disturbance by the observation, it could not be known through which slit an individual particle went.[2] It then would be considered a reasonable supplementation of observable data to say that a percentage of the particles given by $P(A,B_1)$ went through slit B_1, and a percentage given by $P(A,B_2)$ went through slit B_2. When the restrictive interpretation rules this statement out, it does so only

---

[2] The disturbance by the observation, in this case, may be of the same kind as is actually assumed in quantum mechanics: If we observe a particle passing through a slit, it will be pushed off its path by the light ray. It is true that if in this case the particle is not observed at the slit where the observation is made, we know that it passed through the other slit; therefore we have here a knowledge of the particle's position without having disturbed its path. But this knowledge is acquired by inference, not by observation, since what is observed is the absence of the particle at one slit, not its passage at the other. Our fictitious experiment therefore represents a case where, although the observation disturbs, a normal supplementation of observable data by interpolation can be given. This is possible because, in this case, the interference pattern on the screen is of such a kind that it permits us to consider the motion of the particle independent of what happens at a slit through which the particle did not pass.

because the experiment shows that the interference pattern on the screen is not a superposition of the two individual patterns and thus makes it necessary to assume that the path of a particle passing through one slit will depend on whether the other slit is open or not; i.e., it does so because of the causal anomalies derivable from the statement. We often read that the principles of empiricism laid down in the verifiability theory of meaning lead to the rule of restriction formulated by Bohr and Heisenberg. This argumentation is incorrect. Meaning is a matter of definition, and various definitions of meaning can be given; all that can be asked by the philosopher is: Which are the consequences to which a given definition of meaning leads?[3] The restricted meaning of Bohr and Heisenberg's interpretation has the advantage that it eliminates causal anomalies; this is a strong argument in its favor, but it is the only argument. The exhaustive interpretations given in the corpuscle and the wave interpretation are equally compatible with the principles of empiricism if they are conceived as being based on definitions.

It must be added that even a restrictive interpretation is not free from definitions. What we call *phenomena* are certainly not immediate objects of observations; they are inferred from observations by indirect methods (cf. p. 21). These inferences contain a definition, and we shall give in § 29 the exact form of this definition. The logical difference between the physics of phenomena and the physics of interphenomena is therefore a matter of degree; the latter contains more definitions than the former. It is a question of volitional decision which of these two systems we prefer; none can be said to be completely restricted to observational data.

Whereas the Bohr-Heisenberg interpretation uses a *restricted meaning*, we can construct a second form of restrictive interpretation which uses a *three-valued logic*. Ordinary logic is written in terms of the two truth values *true* and *false*. To these we shall add, for the purposes of quantum mechanics, a third truth value which we call *indeterminate*. Statements about unobserved entities then are considered as meaningful; but they are neither true nor false, they are indeterminate. This means that it is impossible to verify or falsify such statements.

The interpretation so constructed is superior to the interpretation by a restricted meaning because it possesses a system of rules which makes it possible to connect statements about unobserved entities with statements about observed entities and thus to manipulate all these statements by means of strictly logical operations. It can be shown that owing to these rules statements expressing causal anomalies always will obtain the truth value of indeterminacy, and therefore can never be asserted as true. On the other hand, part of the statements about unobserved entities are even retained as true, or considered as true in a somewhat wider meaning; they cannot be used, however, for the derivation of causal anomalies because the rules of three-valued logic, differing

---

[3] Cf. the author's *Experience and Prediction* (Chicago, 1938), chap. I.

§8. KINDS OF INTERPRETATIONS

from those of two-valued logic, make such derivations impossible. We shall show, for instance, that the statement: "The particle passes either through slit B_1 or B_2" need not be completely abandoned, but can be retained in a somewhat wider sense, whereas a statement about an action at a distance between the slits is not derivable (cf. § 33). Three-valued logic is therefore the adequate form of quantum mechanics once the decision for the use of a restrictive interpretation has been made.

We have, therefore, good reasons to say that the language of quantum mechanics is written in terms of a three-valued logic. We must not forget, however, that the subject matter of a science, without the addition of further qualifications, does not determine a particular form of logic. Quantum mechanics can be constructed in the form of a two-valued logic; this is demonstrated by the existence of exhaustive interpretations. Only when we introduce the postulate that causal anomalies be not derivable are we obliged to turn to a three-valued logic. It is in this form that the structure of the subject matter expresses itself in the structure of its language. When we apply the same postulate to classical physics we arrive at a two-valued logic. The nature of quantum mechanical occurrences is of such a kind that statements of causal anomalies can be eliminated from the domain of true statements only if a three-valued logic is used; this is the form in which we must express the causal structure of the microcosm.

The three-valued language appears adequate to quantum mechanics because the causal anomalies, as formulated in exhaustive interpretations, appear to be superfluous complications; they need not be taken into account so far as predictions of observable phenomena are concerned. In consideration of this fact a restrictive interpretation will appear natural to one who works in quantum mechanical problems. To this logical insight, habit will add its leveling influence; the desire to ask questions transcending the limits of the restrictive interpretation will disappear; and the restrictive form of quantum mechanics may finally seem to answer everything that can be reasonably asked. This attitude will be as much help to the physicist as the above-mentioned switching over from one interpretation to the other to which he resorts in exhaustive interpretations. Similarly, a man in a world in which bullets behaved as unreasonably as electrons, might learn to restrict his questions in such a way as to obtain only reasonable answers. We should not forget, however, that such an attitude, advisable as it may be, means making a virtue of necessity. When we exclude some kinds of statements about interphenomena and admit others, we have no other reason to do so than that the first statements lead to causal anomalies, whereas the others do not.

Our final judgment concerning the logical significance of restrictive interpretations can therefore be stated as follows. The peculiar form of the causal structure of the microcosm, visible in the causal anomalies of the exhaustive interpretations, finds a corresponding expression in the rule of restriction, or the existence of indeterminate statements, in the restrictive interpretations. In other words: The physical status of the quantum mechanical world, expressed through a restrictive interpretation, is the same as the status expressed through exhaustive interpretations with causal anomalies which can be transformed away locally. The restrictive interpretations do not *say* the causal anomalies, but they do not *remove* them.

The causal anomalies cannot be removed because they are inherent in the nature of the physical world. The *principle of indeterminacy* formulates only one part of this nature; it states that it is impossible to *verify* certain statements about interphenomena. To this is added, by the system of quantum mechanics, another principle which we have called the *principle of anomaly*. It states that no definition of interphenomena can be given which satisfies the requirement of a normal causality; it therefore maintains the impossibility of a normal supplementation of the world of phenomena by interpolation. This includes the restrictive interpretations, since they do not establish a normal causality either.

The limitations of scientific interpretations of the world of quanta, expressed in these two principles, must not be considered as limitations of the power of the human intellect. It is not human ignorance, nor lack of knowledge, which leads to the conditions imposed upon descriptions of the physical world expressed in the laws of quantum mechanics. It is positive knowledge, deep insight into the nature of the atomic world, which constitutes the basis of this strange network of rules, formulated as rules limiting descriptions, but expressing implicitly rules holding for all physical occurrences. Beneath the disguise of a theory of physical knowledge we discern the outlines of a physical world different from what centuries of scientific research had dreamed it to be, but nevertheless demanding recognition as the world of reality.

Part II

OUTLINES OF THE MATHEMATICS OF QUANTUM MECHANICS

§ 9. Expansion of a Function in Terms of an Orthogonal Set

The mathematical formalism of quantum mechanics is based on the general mathematical method of expanding a function in terms of a set of other functions, which are called the *basic functions* of the expansion.

The functions considered are *complex functions of real variables*, i.e., whereas the special values of the argument variables are real numbers, the functions have complex numbers as their special values. We denote complex entities by small Greek letters, real entities by small Latin letters. The complex conjugate of an entity ψ is indicated by ψ^*; the square $|\psi|^2$ of a complex value is the same as $\psi \cdot \psi^*$. The relations to be presented hold, of course, also for real functions; such functions result from the general case by the condition $\psi = \psi^*$. Let us add the remark that the complex functions used in quantum mechanics are assumed to satisfy certain conditions of regularity, and that the infinite series used in the expansions must fulfill some requirement of convergence. We shall not state these conditions here explicitly, leaving this task to the more technical expositions of the subject. It will be the aim of our presentation to point out only the general relations holding for expansions.

A function $\psi(x)$ *is expanded in terms of the set* $\varphi_i(x)$ within the domain D of x, if within D the relation holds

$$\psi(x) = \sum_i \sigma_i \varphi_i(x) \tag{1}$$

The summation runs from $i = 1$ to $i = \infty$; since this is the same in all the following formulae we do not indicate it in the notation. The *expansion coefficients* σ_i are constants, i.e., they do not depend on x; for every function $\varphi_i(x)$ there is such a particular constant σ_i.

The set $\varphi_i(x)$ is called *complete* if the expansion (1) can be given, by means of suitable coefficients σ_i, for every function $\psi(x)$ within D. In addition, the set $\varphi_i(x)$ is usually required to satisfy the further condition that the two relations hold:

$$\sigma_i = \int \psi(x) \varphi_i^*(x) dx \tag{2}$$

$$\int |\psi(x)|^2 dx = \sum_i |\sigma_i|^2 \tag{3}$$

The integration in (2) extends over the domain D; this is the same in all following formulae and therefore does not need a special expression in the notation. Relation (2) determines the expansion coefficients σ_i in terms of the given function $\psi(x)$ and the basic functions $\varphi_i(x)$. This relation shows a structure similar to that of (1), with the difference that here an integration replaces the summation. The symmetry between $\psi(x)$ and σ_i finds a further expression in (3). A set of functions $\varphi_i(x)$ which, for every pair of functions $\psi(x)$ and σ_i connected by (1), satisfies the relations (2) and (3) is called *orthogonal* and *normalized*.

It is understood, in this definition, that the expression (2) is *unique*, i.e., that there is only one set of coefficients σ_i coordinated to a given function $\psi(x)$. This includes the consequence, that, if $\psi(x) = 0$ throughout the domain D, all σ_i must be $= 0$. The normalization finds its expression in (3). For the case that the domain D is infinite, (3) includes the condition that the integral on the left hand side is finite; this condition of convergence is one of the conditions restricting the choice of functions ψ. We say that $\psi(x)$ must be *quadratically integrable*. If, in particular, the value of the quadratical integral in (3) is $=1$, we say also that $\psi(x)$ is normalized. Similarly, (3) imposes upon the σ_i a condition of convergence. Sometimes functions ψ must be considered for which the integral on the left hand side of (3) is not finite, and which therefore cannot be normalized. It is not necessary to deal with such functions in this book.

In (2) and (3) we have given, for the orthogonal and normalized character of the set $\varphi_i(x)$, an *indirect* formulation which makes use of other functions $\psi(x)$ and σ_i. The question arises whether the conditions of orthogonality and normalization of the $\varphi_i(x)$ can be expressed by a *direct* formulation, i.e., in terms of these functions alone, without reference to other functions. This aim can be easily reached when the expansions occurring admit a commutation of summations and integrations. In this case, the orthogonality and normalization of the set $\varphi_i(x)$ can be defined by the conditions:

$$\int \varphi_i(x)\varphi_k^*(x)dx = \delta_{ik} \tag{4}$$

The Weierstrass symbol δ_{ik} used here has the meaning:

$$\delta_{ik} = \begin{cases} 1 \text{ for } i = k \\ 0 \text{ for } i \neq k \end{cases} \tag{5}$$

It is easy to show that, if (1) and (4) hold, the relations (2) and (3) are

§9. EXPANSION OF A FUNCTION

derivable. We prove (2) by multiplying both sides of (1) by $\varphi_k^*(x)$ and integrating with respect to x:

$$\int \psi(x)\varphi_k^*(x)dx = \int \varphi_k^*(x) \cdot \sum_i \sigma_i \varphi_i(x)dx$$

$$= \sum_i \sigma_i \int \varphi_i(x)\varphi_k^*(x)dx \qquad (6)$$

$$= \sum_i \sigma_i \delta_{ik} = \sigma_k$$

Similarly, (3) is proved as follows:

$$\int |\psi(x)|^2 dx = \int \psi(x) \cdot \psi^*(x)dx = \int \sum_i \sigma_i \varphi_i(x) \cdot \sum_k \sigma_k^* \varphi_k^*(x)dx$$

$$= \sum_i \sum_k \sigma_i \sigma_k^* \int \varphi_i(x)\varphi_k^*(x)dx = \sum_i \sum_k \sigma_i \sigma_k^* \delta_{ik} = \sum_i |\sigma_i|^2 \qquad (7)$$

Relation (4) is therefore a *sufficient* condition[1] of orthogonality and normalization in case summations and integrations are commutative. We may ask in which case it is a *necessary* condition; i.e., we may ask what requirement must be made with respect to (1)–(3) in order that (4) be derivable. The answer is that it must be required that the functions $\varphi_i(x)$ are themselves functions to which the expansion can be applied, i.e., functions which can be put in the place of $\psi(x)$. This is proved as follows. If we choose, for instance, a particular function $\varphi_i(x)$ as the function $\psi(x)$, we shall have a solution by putting $\sigma_i = 1$, $\sigma_k = 0$ for $k \neq i$. Since we assumed the expansion to be unique, this must be the only solution. Using (2) and substituting there $\varphi_i(x)$ for $\psi(x)$, we have

$$\left.\begin{aligned} 1 &= \int \varphi_i(x)\varphi_i^*(x)dx \\ 0 &= \int \varphi_i(x)\varphi_k^*(x)dx \qquad k \neq i \end{aligned}\right\} \qquad (8)$$

We thus obtain the relations (4) if the substitution $\varphi_i(x)$ for $\psi(x)$ is made, successively, for every subscript i.

A set of basic functions of this kind may be called *reflexive*, since it includes the basic functions among the functions to be expanded in terms of the set. As it can be shown that the basic functions of (1)–(3) must also be quadratically integrable, it follows that if a set of this type is complete it must also

---
[1] (4) is sufficient to derive (2) and (3) from (1), as shown. If, inversely, (1) and (3) are to be derived from (2), relation (4) cannot be used; instead, a condition corresponding to (16) would have to be introduced, which is combined with the difficulties existing for continuous variables.

be reflexive. Therefore (4) is a consequence of completeness. On the other hand, even if summations and integrations are not commutative, it is possible to derive (2) and (3) from (1) and (4) by more complicated methods, provided that the convergence of the summation in (1) is suitably defined. In an expansion of the type (1), therefore, the indirect characterization of orthogonality can be satisfactorily replaced by a direct characterization.

The definition of completeness given above is also an indirect characterization. Its replacement by a direct characterization represents a rather difficult mathematical problem into which we shall not enter here. Condition (4), in any case, is not sufficient to guarantee completeness. This can be shown as follows. When we omit one of the functions $\varphi_i(x)$, say the function $\varphi_1(x)$, the remaining set will satisfy (4); but it will be impossible to expand into this set a function which, expanded into the original set, has a $\sigma_1 \neq 0$. For, if this were the case, we should have two series, with different coefficients σ_1, for the expansion of the function into the original set, with $\sigma_1 = 0$ in the second series. This shows that completeness represents a specific condition not included in orthogonality and normalization.

The determination whether a certain given set of functions is complete, orthogonal, and normalized, requires special analysis. It can be shown, for instance, that the trigonometrical functions $\varphi_n(x) = \dfrac{1}{\sqrt{2\pi}} e^{inx}$ satisfy these conditions for whole numbers n within the domain D extending from 0 to 2π. These functions are known from the Fourier expansion. Other sets $\varphi_i(x)$ obtain when the $\varphi_i(x)$ are defined as the solutions of certain differential equations; they then are called the *eigen-functions* of these equations. We shall deal with such sets later.

The method of expansion can be extended to functions of a different type. Thus, we can consider *discrete* functions ψ_k ($k = 1, 2, 3 \ldots$) which consist of series of discrete numbers; they are called functions because we can consider the value ψ_k as a function of the subscript. Instead of the functions $\varphi_i(x)$ we then use a set φ_{ik} consisting of a discrete *matrix* of numbers, i.e., a totality of numbers arranged in rows and columns. The matrix will in general be infinite in both directions. The expansion then assumes the form

$$\psi_k = \sum_i \sigma_i \varphi_{ik} \qquad (9)$$

$$\sigma_i = \sum_k \psi_k \varphi_{ik}^* \qquad (10)$$

$$\sum_k |\psi_k|^2 = \sum_i |\sigma_i|^2 \qquad (11)$$

§9. EXPANSION OF A FUNCTION

The conditions of orthogonality and normalization can be expressed by analogy with (4) in the form:

$$\sum_k \varphi_{ik}\varphi_{mk}^* = \delta_{im} \qquad \sum_i \varphi_{ik}\varphi_{in}^* = \delta_{kn} \qquad (12)$$

We have here complete symmetry between ψ_k and σ_i because both are discrete functions. This is the reason that we have here two forms of the condition of orthogonality and normalization. If the summations are commutative, the first form is used to derive (10) and (11) from (9); the second form is used, together with the same assumption, in order to derive (9) and (10) from (11). These proofs are easily given by analogy with (6) and (7). In this discrete case, however, it is possible to eliminate the assumption of commutative summations. The two forms given in (12) represent independent conditions, the combination of which constitutes a very strong condition: It can be generally shown that if both forms (12) hold, the system φ_{ik} is orthogonal, normalized, and even complete, with respect to all functions ψ_k possessing a finite amount of $\sum_k |\psi_k|^2$. This includes a proof that the φ_{ik} are reflexive, since the finite amount of $\sum_i |\varphi_{ik}|^2$ follows from the second form (12) for $k = n$. In the discrete case we therefore have a rather simple form for a direct definition of the three fundamental properties, a definition which is not limited to commutative summations. The proof of this general theorem cannot be given in this book; we refer the reader to the mathematical textbooks on the subject.

Another type of expansion obtains when the functions $\varphi_i(x)$ are replaced by functions of two variables, $\varphi(y,x)$, i.e., when the subscript i is replaced by the variable y. The function $\varphi(y,x)$ may be considered as a *continuous matrix*, and we speak here of the *continuous case* of an expansion. The expansion assumes the form:

$$\psi(x) = \int \sigma(y)\varphi(y,x)dy \qquad (13)$$

$$\sigma(y) = \int \psi(x)\varphi^*(y,x)dx \qquad (14)$$

$$\int |\psi(x)|^2 dx = \int |\sigma(y)|^2 dy \qquad (15)$$

An expansion of this kind is used, for instance, for Fourier expansions when the domain D is infinite; the Fourier functions then have the form

$$\varphi(y,x) = \frac{1}{\sqrt{2\pi}} e^{iyx}.$$

The continuous and the discrete case are *homogeneous* cases of expansions; the first case may be called the *heterogeneous* case. It turns out, however, that

50 PART II. MATHEMATICAL OUTLINES

the three cases are not completely analogous to each other; the continuous case shows some peculiarities. They concern, in particular, the conditions of orthogonality and normalization.

A set of continuous basic functions is never reflexive; therefore relation (4) is here not derivable. Furthermore, the integrations are not commutative, since commutation leads here to indeterminate values. When we wish, in spite of that, to express orthogonality and normalization directly, by a relation analogous to (4), it can be done in a fictitious form. We then write

$$\int \varphi(y,x)\varphi^*(z,x)dx = \delta(y,z) \qquad \int \varphi(y,x)\varphi^*(y,z)dy = \delta(x,z) \qquad (16)$$

The symbol $\delta(x,z)$, which is the continuous analogue of the symbol δ_{ik}, is defined by the conditions

$$\delta(x,z) = 0 \text{ for } x \neq z \qquad \int \delta(x,z)dx = 1 \qquad \int \delta(x,z)dz = 1 \qquad (17)$$

But the symbol has only a fictitious meaning, since no function can be constructed which has these properties. The use of such a fictitious symbol is permissible if a set of rules is given which translate formulae containing the symbol into ordinary formulae. If, in addition, we were given a set of rules for the manipulation of the symbol such that the validity of results derived by means of the symbol were guaranteed, the symbol would have a legitimate place in mathematics. It has so far not been possible, however, to give such rules in general; the symbol, therefore, must be "handled with care" in mathematical derivations.

That the symbol $\delta(x,z)$ must have the properties (17) results from the fact that relations (16) can be derived only by means of an impermissible commutation of integrations. Substituting the value (14) for $\sigma(y)$ in (13) we obtain, when we write z for x in (14):

$$\psi(x) = \iint \psi(z)\varphi(y,x)\varphi^*(y,z)dzdy \qquad (18)$$

Changing the order of integrations, we arrive at:

$$\psi(x) = \int \psi(z)\delta(x,z)dz \qquad (19)$$

$$\delta(x,z) = \int \varphi(y,x)\varphi^*(y,z)dy \qquad (20)$$

Let us first consider the symbol $\delta(x,z)$ used here as an abbreviation whose meaning is given by (20). From (19) we see that the symbol has the property that, multiplied by a function $\psi(z)$ and integrated over z, it reproduces the function ψ. This, as can be easily seen, is precisely the property defined by the conditions (17). Notwithstanding the incorrect way of the derivation of (19),

§9. EXPANSION OF A FUNCTION

the symbol $\delta(x,z)$ in (20) can therefore be identified with the $\delta(x,z)$ symbol of (17), if used for the purpose of certain integrations.

The $\delta(x,z)$-function is called the *Dirac function*, having been introduced by Dirac. It is used in quantum mechanics not only to formulate the condition of orthogonality, but also for various other purposes. It is a fictitious symbol because the conditions (17) cannot be directly realized; they have a meaning only in the sense of approximations. We can define functions $\delta_n(x,z)$ which satisfy (17) approximately; for a given value z they are $= 0$ except when z is within a small interval Δx, within which δ_n represents a steep and high peak. As n increases, this interval becomes narrower, and the height of the peak grows towards infinite values; the integral over x, however, is always $= 1$. The limiting case $\Delta x = 0$ is a degenerate case, and it appears understandable that the use of the Dirac function is criticized by mathematicians. The use of the $\delta(x,z)$-symbol has only a psychological justification: The symbol suggests correct solutions which later can be derived strictly without the use of the symbol.

That for continuous matrices commutation of integrations is not permissible is shown by the continuous Fourier functions $\varphi(y,x) = \dfrac{1}{\sqrt{2\pi}} e^{iyx}$. These functions do not satisfy (16); the integral occurring in these relations is here an indefinite expression and does not vanish for $x \neq z$. For continuous basic functions we therefore prefer to define orthogonality and normalization indirectly by the use of relations (13)–(15). Thus it can be shown that the continuous Fourier functions are normalized and orthogonal in this sense. We may add the remark that in certain cases, for instance for the latter functions, a direct formulation of orthogonality and normalization also can be given in a rigorous way, without the use of the δ-symbol; but the presentation of such methods goes beyond the frame of this book. Let it be said in general that all results of quantum mechanics can be derived by rigorous methods. It may suffice for us to avoid the δ-symbol whenever this can be easily done, and we may be excused when we shall use this symbol occasionally in order to simplify our presentation.

A further peculiarity of the continuous case concerns the changing of variables. Such a change will introduce a density function r into the expansion. If, for instance, we replace y by a variable u such that

$$y = y(u) \qquad dy = \frac{\partial y}{\partial u} du \tag{21}$$

and put

$$r(u) = \frac{\partial y}{\partial u} \tag{22}$$

we arrive at the expansion

$$\psi(x) = \int \sigma(u)\varphi(u,x)r(u)du \tag{23}$$

The density function $r(u)$ must also be introduced into the relations of the δ-symbol. Thus we have, instead of the first integral in (17),

$$\int \delta^{(r)}(u,z) r(u) du = 1 \qquad (24)$$

where $\delta^{(r)}(u,z)$ is to be conceived as the function resulting from $\delta(y,z)$ by the transformation (21). Relations (16) and (14) then remain unchanged, but (15) assumes the form

$$\int |\psi(x)|^2 dx = \int |\sigma(u)|^2 r(u) du \qquad (25)$$

Similar changes result if we replace x by another variable.

§ 10. Geometrical Interpretation in the Function Space

The discrete case lends itself to a geometrical interpretation. Let us assume for a moment that the subscripts run only from 1 to 3, and that ψ_k and σ_i are real numbers. Then both ψ and σ can be considered as vectors in a three-dimensional space, having the three components, respectively, ψ_1, ψ_2, ψ_3, and $\sigma_1, \sigma_2, \sigma_3$. The relation (6), § 9, then represents a transformation of vectors, given by the matrix φ_{ik}.

The structure of a space is determined by its metric. The space which we consider here is *Euclidean*, i.e., its metric is given, for orthogonal coordinates, by the theorem of Pythagoras, which determines the square of the *length* of a vector a as the sum of the squares of its components:

$$a^2 = a_1^2 + a_2^2 + a_3^2 \qquad (1)$$

A generalization of this expression, constructed for two vectors, is the *inner product* (also called *scalar product*):

$$(a,b) = a_1 b_1 + a_2 b_2 + a_3 b_3 \qquad (2)$$

This mathematical expression is used to determine the relation of *orthogonality*. Two non-zero vectors are perpendicular to each other if, and only if, their inner product vanishes. The length can be regarded as a special case of an inner product, namely, as the inner product of a vector with itself.

The orthogonal coordinates which scaffold the space can be conceived as being determined by a set of unit vectors $u_{(1)}, u_{(2)}, u_{(3)}$, which are orthogonal to each other. The vector $u_{(1)}$ has the components 1, 0, 0; the vector $u_{(2)}$ has the components 0, 1, 0; and similarly the vector $u_{(3)}$ has the components 0, 0, 1. The orthogonality and the unit character of this set is expressed by the condition

$$(u_{(i)}, u_{(k)}) = \delta_{ik} \qquad (3)$$

§ 10. GEOMETRICAL INTERPRETATION

A vector a can then be expressed as a linear function of the unit vectors:

$$a = a_1 \cdot u_{(1)} + a_2 \cdot u_{(2)} + a_3 \cdot u_{(3)} \tag{4}$$

Taking the inner product, on both sides, with $u_{(1)}$, and using (3), we obtain

$$(a, u_{(1)}) = a_1 \tag{5}$$

Since a similar relation can be derived for the other components, we can write down the general expression determining the components of a vector in terms of the unit vectors:

$$(a, u_{(i)}) = a_i \tag{6}$$

The inner product is a linear function of the vectors; i.e., it satisfies the two relations

$$(a, k \cdot b) = k \cdot (a, b) \tag{7}$$

$$(a, b + c) = (a, b) + (a, c) \tag{8}$$

where k represents any real number, and a, b, and c any real vectors.

When we wish to introduce, instead of the set $u_{(i)}$, another set $v_{(k)}$ of unit vectors, this is done by the transformation

$$v_{(k)} = \sum_i c_{ki} u_{(i)} \tag{9}$$

The subscribt of summation, here and in the following formulae, runs from 1 to 3. In order to make the set $v_{(k)}$ orthogonal, the coefficients c_{ki} must satisfy the condition of orthogonality

$$\sum_k c_{ki} c_{km} = \delta_{im} \qquad \sum_i c_{ki} c_{mi} = \delta_{km} \tag{10}$$

The orthogonality of the $v_{(k)}$ then follows by the relations

$$(v_{(k)}, v_{(m)}) = \left(\sum_i c_{ki} u_{(i)}, \sum_n c_{mn} u_{(n)} \right) = \sum_i \sum_n c_{ki} c_{mn} (u_{(i)}, u_{(n)})$$

$$= \sum_i \sum_n c_{ki} c_{mn} \delta_{in} = \sum_i c_{ki} c_{mi} = \delta_{km} \tag{11}$$

The reverse transformation to (9) is given by

$$u_{(i)} = \sum_k c_{ki} v_{(k)} \tag{12}$$

This follows with the use of (10) when we multiply (9) on both sides by c_{km}, sum up on k, and finally replace the subscript m by i:

$$\sum_k c_{km} v_{(k)} = \sum_k \sum_i c_{ki} c_{km} u_{(i)} = \sum_i u_{(i)} \sum_k c_{ki} c_{km} = \sum_i \delta_{im} u_{(i)} = u_{(m)} \tag{13}$$

The components of the vector a in the new system of reference are given by

$$a_k' = (a, v_{(k)}) = \left(a, \sum_i c_{ki} u_{(i)}\right)$$

$$= \sum_i c_{ki}(a, u_{(i)}) = \sum_i c_{ki} a_i \qquad (14)$$

This shows that the components of a vector follow the same transformation as the unit vectors. The same holds for the inverse transformation:

$$a_i = \sum_k c_{ki} a_k' \qquad (15)$$

which results from (14) by a proof analogous to (13).

Both length and inner product are *invariants* of orthogonal transformations, i.e., the relations hold:

$$\left.\begin{aligned}(a,a) &= a_1{}^2 + a_2{}^2 + a_3{}^2 = a_1'{}^2 + a_2'{}^2 + a_3'{}^2 \\ (a,b) &= a_1 b_1 + a_2 b_2 + a_3 b_3 = a_1' b_1' + a_2' b_2' + a_3' b_3'\end{aligned}\right\} \qquad (16)$$

It is sufficient to show this for the inner product, since the length is a special case of such a product. We have

$$(a,b) = \sum_i a_i b_i = \sum_i \sum_k \sum_m c_{ki} a_k' c_{mi} b_m' = \sum_k \sum_m \left(\sum_i c_{ki} c_{mi}\right) a_k' b_m'$$

$$= \sum_k \sum_m \delta_{km} a_k' b_m' = \sum_k a_k' b_k' \qquad (17)$$

All these relations can be extended to an infinite number of dimensions, if only precautions are taken concerning the convergence of expressions like the length and the inner product; the components a_i of a vector then are assumed to converge toward zero with growing subscript i in such a way that these expressions remain finite. Similar precautions permit us to consider summations as commutative. With these qualifications the formulae presented hold equally when the subscript of summation runs from 1 to ∞.

Furthermore, the relations can be generalized for the case of complex vectors, i.e., vectors whose components are complex numbers. We then define the inner product by the expression

$$(\psi,\sigma) = \sum_i \psi_i \cdot \sigma_i^* \qquad (18)$$

and similarly the square of the length by

$$(\psi,\psi) = \sum_i \psi_i \cdot \psi_i^* \qquad (19)$$

§ 10. GEOMETRICAL INTERPRETATION

It follows that the length is a positive real number, which vanishes only if ψ vanishes. The inner product is, in general, a complex number, which satisfies the relation

$$(\psi,\sigma) = (\sigma,\psi)^* \qquad (20)$$

and, in addition, has the character of linearity as expressed in (7)–(8):

$$\left.\begin{aligned}(\kappa\cdot\psi,\sigma) &= \kappa \cdot (\psi,\sigma) \\ (\psi,\kappa\cdot\sigma) &= \kappa^* \cdot (\psi,\sigma) \\ (\psi,\sigma + \tau) &= (\psi,\sigma) + (\psi,\tau)\end{aligned}\right\} \qquad (21)$$

Here κ is a complex constant, not a vector. For real numbers all these relations are identical with the above given. The extension to complex numbers must be considered as a convention enabling us to handle complex vectors by analogy with real vectors. Thus we shall say that two complex vectors are orthogonal to each other when the inner product (18) vanishes.

Combining the two forms of generalization, we shall consider the formulae (18)–(21) as holding also for the case of an infinite number of dimensions, i.e., for subscripts i running from 1 to ∞. The existence of the quantities (18) and (19) is guaranteed by the definition of the vector; i.e., entities ψ and σ for which length and inner product do not possess determinate finite values, are not called vectors.

The space defined by the metric (18) and (19) is called *unitary space*, or *Hilbert space*. This term denotes, for an unrestricted number of dimensions, the complex analogue of Euclidean space, namely, a space in which a system of orthogonal straight-line coordinates is possible. Introducing complex unit vectors $\epsilon_{(i)}$ satisfying the relations of orthogonality

$$(\epsilon_{(i)}, \epsilon_{(k)}) = \delta_{ik} \qquad (22)$$

we can express every vector ψ as a linear function of the unit vectors:

$$\psi = \sum_k \psi_k \epsilon_{(k)} \qquad (23)$$

As before, the components ψ_k are given by the relation

$$\psi_k = (\psi,\epsilon_{(k)}) \qquad (24)$$

A transformation to another set of orthogonal unit vectors $\eta_{(i)}$ is called a *unitary transformation* and is given by

$$\eta_{(i)} = \sum_k \epsilon_{(k)}\, \varphi_{ik} \qquad \epsilon_{(k)} = \sum_i \eta_{(i)} \varphi_{ik}^* \qquad (25)$$

where the coefficients satisfy the relation of orthogonality

$$\sum_k \varphi_{ik}\varphi_{mk}^* = \delta_{im} \qquad \sum_i \varphi_{ik}\varphi_{in}^* = \delta_{kn} \qquad (26)$$

By a derivation analogous to (13) we can show that the second of the relations (25) follows from the first, when the relation (26) are assumed. The proof, that owing to these conditions the vectors $\eta_{(i)}$ have the length 1 and are orthogonal to each other, is easily given by analogy with (11). The same coefficients φ_{ik} determine the transformation of the components ψ_k of the vector ψ into new components σ_i:

$$\psi_k = \sum_i \sigma_i \varphi_{ik} \qquad \sigma_i = \sum_k \psi_k \varphi_{ik}^* \qquad (27)$$

These relations are proved by analogy with (14). Length and inner product are invariants of such a transformation; i.e., the relations hold:

$$(\psi,\psi) = \sum_k \psi_k \psi_k^* = \sum_i \sigma_i \sigma_i^* \qquad (28)$$

$$(\psi,\chi) = \sum_k \psi_k \chi_k^* = \sum_i \sigma_i \tau_i^* \qquad (29)$$

if χ is a second vector in the ϵ-system and the τ_i represent its components in the η-system.

Relations (26) and (27) correspond to the relations (12), § 9, and (9)–(10), § 9; (28) corresponds to (11), § 9. We see that the expansion into a set of orthogonal functions corresponds to a unitary transformation in a unitary space of an infinite number of dimensions.

According to a remark by F. Klein[1] any such transformation can be interpreted in two ways, which may be distinguished as the *passive* and the *active* interpretation. In the passive interpretation, which we have used so far, we consider ψ and σ as the *same* vector, but represented in *different* systems of coordinates. The matrix φ_{ik} then represents a *transformation of coordinates*, and (9), § 9, determines the components of the vector in the ψ-system as functions of the components of the same vector in the σ-system. Since the transformation is unitary, it consists in a mere turning of the system of coordinates, combined with mirror reflections.

In the active interpretation we consider ψ and σ as *different* vectors, but represented in the *same* system of coordinates. The matrix φ_{ik} then represents an *operator*. An operator, in the general sense of the term, is a rule which coordinates to a given entity, or function, another entity, or function. Such a coordination in terms of an operator can always be interpreted as a deformation of the space, which transforms the one entity, or function, into the other. In the case considered here the operator φ_{ik} coordinates to every given vector another vector. Owing to the unitary character of the transformation, the deformation of the space produced by it is of a rather simple kind: it consists merely in a turning of the space, combined with mirror reflections. In another

---
[1] F. Klein, *Elementarmathematik vom höheren Standpunkte aus*, Vol. II (Berlin, 1925), p. 74.

§ 10. GEOMETRICAL INTERPRETATION

version of the active interpretation we speak, instead of a deformation of the space, of a coordination of two different spaces to each other performed by the matrix φ_{ik}.

The geometrical language can be extended to the continuous case (13)–(15), § 9. The functions ψ and σ then are considered as vectors in a space F whose number of dimensions has the magnitude of the continuum. We must realize that, in the space F, x and y are not coordinates; they rather perform the numeration of dimensions, such that a special value of x determines a dimension, as does the subscript i for discrete functions, whereas the value of the coordinate in this dimension is the value $\psi(x)$. The totality of all values $\psi(x)$ is the vector ψ. The space F is therefore called the *function space*, since each of its points corresponds to a whole function. As before, the metric of the space is determined by the expressions introduced for length and inner product, the existence of which represents a condition defining vectors. These expressions are given here by the relations:

$$(\psi,\psi) = \int \psi(x)\psi^*(x)dx = \int \sigma(y)\sigma^*(y)dy = (\sigma,\sigma) \qquad (30)$$

$$(\psi,\chi) = \int \psi(x)\chi^*(x)dx = \int \sigma(y)\tau^*(y)dy = (\sigma,\tau) \qquad (31)$$

Relation (30) corresponds to (15), § 9. Unitary transformations can be defined either by the fictitious way of writing used in (16), § 9, or by the requirement that the expressions (30) and (31) be invariants of the transformation. In the passive interpretation of the transformation, the axes determined by the various values of x are different from the axes determined by the various values of y. In the active interpretation these axes can be identified.

The geometrical language can also be used for the heterogeneous case. Using the passive interpretation we then consider $\psi(x)$ and σ_i as representations of the same vector in different systems of coordinates, the transformation between these systems being given by the heterogeneous matrix $\varphi_i(x)$. The invariance of the length is expressed by (3), § 9. It follows that in this interpretation the same space can be built up either by a denumerable infinity or by a continuous infinity of dimensions.[2] In the active interpretation of this case we use the version of different spaces. The mixed matrix $\varphi_i(x)$ then coordinates the vectors of a space of a denumerable infinity of dimensions to those of a space of a continuous infinity of dimensions.

---

[2] By the term "number of dimensions" we understand here the number of parameters determining a point in the space. Another meaning of the term is used when it is defined by the number of linearly independent vectors, which is denumerable also for a space having, in the first terminology, a continuous number of dimensions. Only for a finite number of dimensions do both meanings coincide.

§ 11. Reversion and Iteration of Transformations

The geometrical language enables us to discuss mathematical operations with orthogonal expansions in terms of unitary transformations. In order to analyze the nature of transformations it is necessary to answer two questions: first, which is the converse of a given transformation, and second, which is the transformation resulting from the iteration of two transformations. In answering these questions for unitary transformations, we shall develop further properties of orthogonal expansions. We begin with the discrete case.

The question of the *reversed transformation* is easily answered by (10), § 9. Let φ^{-1} be the *converse* of φ; by analogy with (9), § 9, we then define φ^{-1} by the relation

$$\sigma_i = \sum_k \psi_k \varphi_{ki}^{-1} \tag{1}$$

This definition is chosen in such a way that, as in (9), § 9, the summation runs over the first subscript of φ_{ki}^{-1}. The choice of the letters i and k is of course irrelevant. Now a comparison with (10), § 9, shows that

$$\varphi_{ki}^{-1} = \varphi_{ik}^* = \breve{\varphi}_{ki} \tag{2}$$

This means that the reversed transformation obtains by reversing the subscripts and taking the complex conjugate of the original transformation. The matrix thus resulting, which is called the *adjoint matrix* to φ_{ki}, will be denoted by the arc symbol ($\smile$).[1] We see that in the theory of unitary transformations the symbol φ^{-1} is dispensable. Since we are concerned here only with such transformations, we shall consequently always express the reversed transformation immediately by the symbol $\breve{\varphi}$. Whereas we derive here condition (2) from the definition of unitary transformations by the condition of orthogonality (7), § 9, condition (2) is frequently used, inversely, as the definition of unitary transformations.

The question of iteration of transformations is to be asked as follows. We have considered so far a transformation between the ψ-system and the σ-system, given by the matrix φ_{ik}. Let us now consider a further transformation between the σ-system and a third system which we call the τ-system, and let ω_{km} be the matrix effecting the transformation between the two latter systems. We then have

$$\psi_k = \sum_i \sigma_i \varphi_{ik} \tag{3}$$

$$\sigma_i = \sum_m \tau_m \omega_{mi} \tag{4}$$

---

[1] Instead of this symbol, the symbol † is used in some other notations.

§11. REVERSION AND ITERATION

From these relations we obtain the direct transformation between the ψ-system and the τ-system as follows:

$$\left.\begin{aligned}\psi_k &= \sum_i \sum_m \tau_m \omega_{mi} \varphi_{ik} \\ &= \sum_m \tau_m \chi_{mk}\end{aligned}\right\} \quad (5)$$

$$\chi_{mk} = \sum_i \omega_{mi} \varphi_{ik} \quad (6)$$

If φ and ω are orthogonal, so is χ. This is shown as follows:

$$\begin{aligned}\sum_k \chi_{mk} \chi_{nk}^* &= \sum_k \sum_i \omega_{mi} \varphi_{ik} \sum_l \omega_{nl}^* \varphi_{lk}^* \\ &= \sum_i \sum_l \omega_{mi} \omega_{nl}^* \sum_k \varphi_{ik} \varphi_{lk}^* \\ &= \sum_i \sum_l \omega_{mi} \omega_{nl}^* \delta_{il} = \sum_i \omega_{mi} \omega_{ni}^* = \delta_{mn}\end{aligned} \quad (7)$$

We therefore can consider (5) as the expansion of ψ_k in the orthogonal set χ_{mk}. The iteration of two orthogonal expansions constitutes a new orthogonal expansion.

The relation (6) determines the new set χ as a function of the two given sets ω and φ; in this relation we meet the *matrix multiplication* of tensors, which is asymmetrical in the subscript of summation. We can solve this relation for either ω_{mi} or φ_{ik} by multiplying, respectively, with φ_{lk}^* or ω_{ml}^*, and summing up on k or m; we thus obtain

$$\omega_{ml} = \sum_k \chi_{mk} \varphi_{lk}^* \quad (8)$$

$$\varphi_{lk} = \sum_m \omega_{ml}^* \chi_{mk} \quad (9)$$

Using the sign of the reversed transformations we can write these relations:

$$\omega_{ml} = \sum_k \chi_{mk} \breve{\varphi}_{kl} \quad (10)$$

$$\varphi_{lk} = \sum_m \breve{\omega}_{lm} \chi_{mk} \quad (11)$$

In this form of writing, these relations, too, assume the form of a matrix multiplication.

These relations can be illustrated by a diagram, as given in figure 7, in which the arrows indicate the transformations. The direction of the arrows is chosen in such a way that, for instance, φ is interpreted as a transformation leading from σ to ψ; this corresponds to (3), since this relation determines ψ for a given σ. The arrow χ is, geometrically speaking, the sum of the arrows ω and φ; the corresponding transformation χ is determined by (6) as the matrix product of the two transformations ω and φ.[2] The general form of the relation (6) is expressed by the following relation in which a transformation like φ is represented by a symbol of the form $Trn(\sigma,\psi)$, and in which the cross expresses matrix multiplication:

$$Trn(\tau,\psi) = Trn(\tau,\sigma) \times Trn(\sigma,\psi) \qquad (12)$$

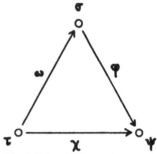

Fig. 7. Triangle of transformations.

The relations (10) and (11) result when we reverse, respectively, the directions of the arrows φ and ω; and they can be expressed in a form corresponding to (12).

The relations holding for the triangle of transformations are summarized in the following formulae:

$$\psi_k = \sum_i \sigma_i \varphi_{ik} = \sum_m \tau_m \chi_{mk} \qquad (13a)$$

$$\sigma_i = \sum_m \tau_m \omega_{mi} = \sum_k \psi_k \breve{\varphi}_{ki} \qquad (13b)$$

$$\tau_m = \sum_k \psi_k \breve{\chi}_{km} = \sum_i \sigma_i \breve{\omega}_{im} \qquad (13c)$$

$$\chi_{mk} = \sum_i \omega_{mi} \varphi_{ik} \qquad (13d)$$

$$\omega_{ml} = \sum_k \chi_{mk} \breve{\varphi}_{kl} \qquad (13e)$$

$$\varphi_{lk} = \sum_m \breve{\omega}_{lm} \chi_{mk} \qquad (13f)$$

Relation (13d) admits of a further interpretation. Let us consider the subscript m as constant, and denote this by putting the m into brackets. (13d) then represents the expansion of $\chi_{(m)k}$ into the orthogonal set φ_{ik}, with $\omega_{(m)i}$ as expansion coefficients. Now $\chi_{(m)k}$ represents, for a variable m, not one function, like ψ_k, but a set of functions each of which has a particular value of m. We see that the set $\omega_{(m)i}$ represents the corresponding expansion coefficients for an

---

[2] In the diagram we use a vectorial representation in which a vector cannot be shifted parallel to itself. This condition is necessary in order to make vectorial addition non-commutative, corresponding to the multiplication of matrices.

§11. REVERSION AND ITERATION

expansion in the φ_{ik}.[3] Similar interpretations can be given for (13e) and (13f). We have written these equations in such a way that the first term on the right hand side corresponds to the expansion coefficients, the succeeding term, to the basic functions. Our result can be stated as follows: In a triangle of transformations we can choose the direction of the transformations in such a way that a given transformation can be considered as the set of expansion coefficients belonging to the expansion of the second transformation in terms of the third.

Our results can be transferred to the continuous case. The reversed transformation $\breve{\varphi}(x,y)$ then is given by:

$$\sigma(y) = \int \psi(x)\breve{\varphi}(x,y)dx \tag{14}$$

$$\breve{\varphi}(x,y) = \varphi^*(y,x) \tag{15}$$

For the triangle of transformations we have here the relations:

$$\psi(x) = \int \sigma(y)\varphi(y,x)dy = \int \tau(z)\chi(z,x)dz \tag{16a}$$

$$\sigma(y) = \int \tau(z)\omega(z,y)dz = \int \psi(x)\breve{\varphi}(x,y)dx \tag{16b}$$

$$\tau(z) = \int \psi(x)\breve{\chi}(x,z)dx = \int \sigma(y)\breve{\omega}(y,z)dy \tag{16c}$$

$$\chi(z,x) = \int \omega(z,y)\varphi(y,x)dy \tag{16d}$$

$$\omega(z,y) = \int \chi(z,x)\breve{\varphi}(x,y)dx \tag{16e}$$

$$\varphi(y,x) = \int \breve{\omega}(y,z)\chi(z,x)dz \tag{16f}$$

These relations correspond exactly to those of the discrete case given in (13).

In the heterogeneous case the situation is more complicated because the step from $\psi(x)$ to σ_i is structurally different from the reversed step, owing to the difference between continuous and discrete variables. We therefore cannot define

---

[3] This interpretation cannot be reversed, i.e., it would be wrong to say that we can also consider ω_{mk} in (13d) as basic functions and φ_{ik} as the set of expansion coefficients. The reason is that our expansions are so defined that the summation runs through the first subscript. We can, however, transform (13d) into the form

$$\chi_{mk} = \sum_i \breve{\varphi}_{ki}{}^* \breve{\omega}_{im}{}^*$$

In this form the function $\breve{\varphi}^*$ represents the expansion coefficients, and $\breve{\omega}^*$ the basic functions.

a reversed transformation which is of the same structure as the original transformation, and must leave the reversed step in the form (2), § 9.

There will be differences, furthermore, in the iteration of transformations according as we choose the transformation ω as a discrete or as a heterogeneous matrix. Using a discrete ω we have for the triangle of transformations:

$$\psi(x) = \sum_i \sigma_i \varphi_i(x) = \sum_m \tau_m \chi_m(x) \tag{17a}$$

$$\sigma_i = \sum_m \tau_m \omega_{mi} = \int \psi(x) \varphi_i^*(x) dx \tag{17b}$$

$$\tau_m = \int \psi(x) \chi_m^*(x) dx = \sum_i \sigma_i \breve{\omega}_{im} \tag{17c}$$

$$\chi_m(x) = \sum_i \omega_{mi} \varphi_i(x) \tag{17d}$$

$$\omega_{mi} = \int \chi_m(x) \varphi_i^*(x) dx \tag{17e}$$

$$\varphi_i(x) = \sum_m \breve{\omega}_{im} \chi_m(x) \tag{17f}$$

These relations correspond to those of the homogeneous cases, with the modification that the symbol $\varphi_i^*(x)$ takes the place of a symbol for the reversed

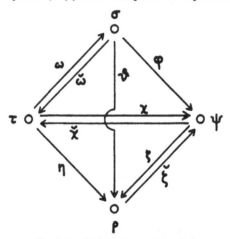

Fig. 8. Quadrangle of transformations.

transformation. As before, these relations are written in such a form that the first term on the right hand side corresponds to the expansion coefficients, the

§11. REVERSION AND ITERATION

succeeding term, to the basic functions. The corresponding relations for a heterogeneous ω can easily be derived.

The given relations can be extended to the case of four different basic functions, for which we construct a quadrangle of transformations (figure 8). We shall write the relations only for the discrete cases. Thus, in figure 8 we have, always following the rules of the arrows and noticing that the use of the arc sign means reversion of the arrow:

$$\varphi_{ik} = \sum_{m} \breve{\omega}_{im}\chi_{mk} \qquad \zeta_{kl} = \sum_{n} \breve{\chi}_{kn}\eta_{nl} \qquad (18)$$

$$\vartheta_{il} = \sum_{k} \varphi_{ik}\zeta_{kl} = \sum_{k}\sum_{m}\sum_{n} \breve{\omega}_{im}\chi_{mk}\breve{\chi}_{kn}\eta_{nl} = \sum_{m} \breve{\omega}_{im}\eta_{ml} \qquad (19)$$

since

$$\sum_{k} \chi_{mk}\breve{\chi}_{kn} = \delta_{kn} \qquad (20)$$

We see that the relations of the triangle of transformations, which we have assumed for the three triangles $\sigma\psi\tau$, $\psi\tau\rho$, $\sigma\psi\rho$, hold also for the fourth triangle $\sigma\rho\tau$. The same can be proved for the continuous and for the heterogeneous case.

We can use these results to establish a relation between the two diagonal transformations ϑ and χ. With

$$\breve{\omega}_{im} = \sum_{k} \varphi_{ik}\breve{\chi}_{km} \qquad (21)$$

(19) leads to

$$\vartheta_{il} = \sum_{k}\sum_{m} \varphi_{ik}\breve{\chi}_{km}\eta_{ml} \qquad (22)$$

In the interpretation of the diagram this means that in the "sum" (19) we have replaced the term $\breve{\omega}_{im}$ by the "sum" (21). Relation (22) therefore states that the train of arrows φ, $\breve{\chi}$, η is equivalent to the arrow ϑ. Applying the form of writing used in (12) we can write (22):

$$Trn(\sigma,\rho) = Trn(\sigma,\psi) \times Trn(\psi,\tau) \times Trn(\tau,\rho) \qquad (23)$$

All these relations are easily transferred to the continuous and to the heterogeneous case. Let us write down for later use only the following relations, which are taken from the quadrangle of transformations:

$$\psi(x) = \sum_{k} \sigma_{k}\varphi_{k}(x) = \sum_{m} \rho_{m}\breve{\zeta}_{m}(x) \qquad (24)$$

$$\rho_{m} = \sum_{k} \sigma_{k}\vartheta_{km} \qquad (25)$$

$$\varphi_{k}(x) = \sum_{m} \vartheta_{km}\breve{\zeta}_{m}(x) \qquad (26)$$

$$\chi_{k}(x) = \sum_{m} \eta_{km}\breve{\zeta}_{m}(x) \qquad (27)$$

§ 12. Functions of Several Variables and the Configuration Space

The relations presented can be transferred to the case of functions of several variables. Let us begin with the continuous case (13), § 9. We shall replace x by a set of variables $x_1 \ldots x_n$, and y by a set $y_1 \ldots y_n$; correspondingly the integration is to be replaced by a multiple integration. Thus we have, instead of (13)–(16), § 9:

$$\psi(x_1 \ldots x_n) = \int \ldots \int \sigma(y_1 \ldots y_n) \varphi(y_1 \ldots y_n, x_1 \ldots x_n) dy_1 \ldots dy_n \tag{1}$$

$$\sigma(y_1 \ldots y_n) = \int \ldots \int \psi(x_1 \ldots x_n) \varphi^*(y_1 \ldots y_n, x_1 \ldots x_n) dx_1 \ldots dx_n \tag{2}$$

$$\int \ldots \int |\psi(x_1 \ldots x_n)|^2 dx_1 \ldots dx_n = \int \ldots \int |\sigma(y_1 \ldots y_n)|^2 dy_1 \ldots dy_n \tag{3}$$

$$\int \ldots \int \varphi(y_1 \ldots y_n, x_1 \ldots x_n) \varphi^*(z_1 \ldots z_n, x_1 \ldots x_n) dx_1 \ldots dx_n$$
$$= \delta(y_1, z_1) \ldots \delta(y_n, z_n) \tag{4}$$

The condition of orthogonality and normalization is here expressed in the fictitious form introduced in (16), § 9; we use in this case a product of δ-symbols, which is different from zero only if for each of the symbols the corresponding values y_i and z_i are equal to each other; otherwise this product is $= 0$. All theorems and proofs given for the case of one variable can be similarly transcribed for functions with several variables. The theory of this generalized expansion is therefore formally given with the theory of the simple expansion.

For the heterogeneous case (1), § 9, we can develop similar formulae. Instead of the coefficients σ_i we then shall obtain coefficients $\sigma_{i_1} \ldots \sigma_{i_n}$, and instead of the functions $\varphi_i(x)$ we shall have functions $\varphi_{i_1} \ldots \varphi_{i_n}(x_1 \ldots x_n)$. Since the number of these constants and functions is denumerable, we can, however, introduce a numeration running in one subscript i, and, instead of a multiple summation, we then have simply one summation in i. We therefore obtain the formulae:

$$\psi(x_1 \ldots x_n) = \sum_i \sigma_i \varphi_i(x_1 \ldots x_n) \tag{5}$$

$$\sigma_i = \int \ldots \int \psi(x_1 \ldots x_n) \varphi_i^*(x_1 \ldots x_n) dx_1 \ldots dx_n \tag{6}$$

$$\int \ldots \int |\psi(x_1 \ldots x_n)|^2 dx_1 \ldots dx_n = \sum_i |\sigma_i|^2 \tag{7}$$

$$\int \ldots \int \varphi_i(x_1 \ldots x_n) \varphi_k^*(x_1 \ldots x_n) dx_1 \ldots dx_n = \delta_{ik} \tag{8}$$

§ 12. SEVERAL VARIABLES

The discrete case (9), § 9, need not be generalized, because a discrete function ψ_{ik} can always be replaced by a discrete function in one subscript, ψ_i, at least for the case in which only multiple summations in both subscripts are used.

We now shall consider a second geometrical interpretation different from the one we gave in § 10, which is particularly appropriate for the case of several variables, although it can also be carried through for one variable x. In this interpretation we consider the $n+1$-dimensional space C built up by the n variables $x_1 \ldots x_n$ and, in addition, a dimension for ψ, which is usually called *configuration space*. The expansion (1) then can be considered as a transformation in the space C.

As in the other case, we have here a passive and an active interpretation. In the passive interpretation we consider the functions $\psi(x_1 \ldots x_n)$ and $\sigma(y_1 \ldots y_n)$ as identical; the transformation then is a transformation of coordinates. In the active interpretation we identify the systems $x_1 \ldots x_n$ and $y_1 \ldots y_n$; the transformation then has the character of an operator and performs a coordination between the surfaces ψ and σ. We may here also use the version in which the space of the $x_1 \ldots x_n$ is distinguished from the space $y_1 \ldots y_n$; the transformation then has the same operator interpretation.

We must now explain a fundamental difference between this geometrical interpretation and the interpretation presented in § 10. In the latter interpretation the transformation has the character of a *point transformation*. If we use the active interpretation, this means that each point in the function space F determines a coordinated point in F. The interpretation in terms of the configuration space C, however, does not lead to point transformations. The whole surface σ determines a surface ψ; but this coordination is not done point by point, since a change in the shape of σ within a certain domain changes the value of the integral in (1) and therefore changes the whole surface ψ. We even cannot say which point on ψ is to be coordinated to a given point on σ. We may speak here of a *holistic transformation*, since it is a coordination between the surfaces as wholes which is performed in (1).

The same feature is to be found in the passive interpretation; here it means that the transformation of coordinates is not a point transformation, but a holistic transformation, depending on the shape of the surface ψ (or σ). Since this is different from what is usually called a transformation of coordinates, the active interpretation seems preferable for this case.

Since the relation between ψ and σ in (1) is expressed in terms of an integral, the expansion (1) is also called an *integral transformation*. We see that an integral transformation can be interpreted either as a point transformation in the function space or as a holistic transformation in the configuration space.

For the discrete case and the heterogeneous case the interpretation in the configuration space seems less appropriate, although, of course, it can be carried through.

§ 13. Derivation of Schrödinger's Equation from de Broglie's Principle

In the preceding sections we presented relations belonging entirely to the domain of mathematics. In the present and in the following sections we turn to physics. It is the quantum mechanical method of establishing relations between physical entities which we now must consider.

In order to understand this method we must realize that quantum mechanics is constructed as a generalization of classical mechanics. This generalization has been achieved by the establishment of a rule by means of which an equation of classical mechanics is transcribed into an equation of quantum mechanics. Since classical relations are causal relations, and quantum relations are probability relations, the rule of transcription is so constructed that it determines probability laws by analogy with causal laws. We are bound to use such a method of generalization, because classical mechanics represents the only starting point from which we can construct the new domain of quantum mechanics. On the other hand, it is clear that we have no purely logical directive for the establishment of the method of generalization. All that is to be required logically is that the new relations be identical with classical relations for the limiting case $h = 0$; but this requirement leaves the rules of generalization arbitrary within wide limits.

The way towards the construction of these rules, therefore, could not be found by logical reasoning. It was the instinct of the physicist which pointed the way. It is true that the men who did the work felt obliged to adduce logical reasons for the establishment of their assumptions; and it seems plausible that this apparently logical line of thought was an important tool in the hands of those who were confronted by the task of transforming ingenious guesses into mathematical formulae. L. de Broglie[1] was led by the consideration that the duality of waves and corpuscles discovered for light should be extended to hold equally for the elementary particles of matter; Schrödinger[2] was guided by the analogy of mechanics and optics which enabled him to construct his transition from classical mechanics to wave mechanics on the model of the transition from geometrical optics to wave optics; Heisenberg believed that, since statements about the orbit of an electron inside the atom cannot be directly verified, statements about transition probabilities expressed as relations between matrices must include all that can be said about elementary particles. The analysis, added in a succeeding period of criticism, showed that, although the conclusions of these inferences were true, the inferences themselves could not be considered as valid. What justifies us today in considering the conclusions as a well-founded physical theory is the amazing correspondence of the mathe-

---
[1] *Ann. d. Phys.* (10), Vol. 3 (Paris, 1925), p. 22.
[2] *Ann. d. Phys.* (4), Vol. 79 (Leipzig, 1926), pp. 361, 489.

§13. SCHRÖDINGER'S EQUATION

matical system obtained, with known observational results, in combination with its predictive power manifested in the results of newly devised experiments. The historical development of quantum mechanics, therefore, constitutes an illustration of the distinction between *context of discovery* and *context of justification*, a distinction which must be made for all kinds of scientific inquiry.[3] The path of discovery runs through "series of inferences which are deeply veiled by the darkness of instinctive guessing," if I am permitted to apply here a phrase which Schrödinger once used in a letter to me, written some years before his great quantum mechanical discoveries. Once a theory has been constructed, it is to be judged within the context of justification, i.e., by the inductive evidence conferred to it through its empirical success.

To make this distinction clear let us consider a way in which Schrödinger's differential equation for waves of matter can be derived from the principles introduced by Planck and L. de Broglie. This is not the way actually used by Schrödinger; it is rather a logical short cut of this way, which has been constructed later and which we use because an exact analysis of Schrödinger's ideas would lead us too far away from the purpose of a merely logical analysis with which this book is concerned.

The introduction of particle concepts into the wave theory of light goes back to Planck's introduction of the quantum h. According to his assumption, every light wave of the frequency ν is equipped with an energy quantum of the amount[4]

$$H = h \cdot \nu \qquad (1)$$

Planck's assumption was supplemented by Einstein's idea that, in a similar way, a momentum of the amount

$$p = \frac{h\nu}{c} = \frac{h}{\lambda} \qquad (2)$$

can be coordinated to each light wave. The momentum differs from the energy so far as it is an entity equipped with a direction; therefore it is to be represented by a vector with the components p_i. Similarly we have vectors with the components λ_i and c_i. Relation (2) holds only for the absolute amounts p, c, and λ of the vectors. For the components we have here the relation:

$$p_i = \frac{h}{\lambda^2} \lambda_i \qquad (3)$$

These relations are greatly simplified when we introduce the vector *wave number* with the components

$$b_i = \frac{\lambda_i}{\lambda^2} \qquad (4)$$

---
[3] Cf. the author's *Experience and Prediction* (Chicago, 1938), p. 7.
[4] We follow the usual notation expressing frequency and wave length by the letters "ν" and "λ", thus deviating from our rule of reserving Greek letters for complex values.

The absolute amount b of this vector measures the number of waves in a unit of length in the direction of the wave motion and is given by $b = \dfrac{1}{\lambda}$. Using (4), we can write (2) and (3) in the form

$$p = h \cdot b \qquad p_i = h \cdot b_i \tag{5}$$

The momentum relation is thus made analogous to Planck's relation (1), with the formal difference, however, that the energy relation is a scalar relation, whereas the momentum relation is a vector relation.

Einstein's assumption could be interpreted as endowing light waves with the characteristics of corpuscles, and thus as a statement of a dualistic nature of light. L. de Broglie made the decisive step of extending this dualism of waves and corpuscles to particles of matter by coordinating a wave to each mass particle. The frequency ν of this wave was determined by (1) in terms of the energy of the particle. L. de Broglie saw that the velocity w of these waves must be different from the light velocity c and that, therefore, the momentum of the wave must be written, by analogy with (2) and (5), in the form

$$p = \frac{h\nu}{w} = \frac{h}{\lambda} = h \cdot b \qquad p_i = h \cdot b_i \tag{6}$$

From the relativistic expression for the momentum of a free particle

$$p = \frac{H v}{c^2} = \frac{h\nu v}{c^2} \tag{7}$$

where v is the velocity of the particle, L. de Broglie inferred that for a free particle the velocity of the wave (i.e., the phase velocity) must be given by

$$w = \frac{c^2}{v} \tag{8}$$

In contradistinction to (8), relations (1) and (6) are not restricted to free particles, but hold for particles in all conditions. Let us consider a particle moving in a potential field $U(q_1, q_2, q_3)$, where the q_i as usual denote the position coordinates of the particle. When we now use only the nonrelativistic form the relation which holds between energy and momentum of the particle is given by:

$$H = \frac{1}{2m}(p_1^2 + p_2^2 + p_3^2) + U(q_1, q_2, q_3) \tag{9}$$

The first expression on the right hand side denotes the kinetic energy of the particle of the mass m, and the second expression denotes its potential energy in a field of force $U(q_1, q_2, q_3)$; H is the total energy of the particle.

Now let us first consider a simplified case, namely, the case that the potential U is constant, i.e., does not depend on the q_i. If, in particular, $U = 0$, we have the case of a free particle; the case $U = $ constant, however, is not essen-

§13. SCHRÖDINGER'S EQUATION

tially different from this case. Since the force acting upon a particle is given by the derivative of U, the case $U =$ constant represents a case in which no field of force exists. Let us express the constancy of U by omitting the arguments q_i.

We now can introduce the relations (1) and (5) into the equation (9). We thus arrive at the relation

$$h \cdot \nu = \frac{h^2}{2m}(b_1{}^2 + b_2{}^2 + b_3{}^2) + U \qquad (10)$$

We see that the Planck–de Broglie relations permit us to transcribe an equation holding between energy and momenta into an equation holding between frequency and wave number. In other words: The Planck–de Broglie relations can be used to transcribe a particle equation into a wave equation. The latter may be called a *frequency wave-length equation*, since it connects these entities. It is also called a *law of dispersion*, since it can be transformed into a law connecting frequency and velocity of the waves when the relation $\nu = b_i w_i$ is used.

Now let us assume that the wave is given by the complex function

$$\psi = \psi_0 e^{2\pi i(b_1 q_1 + b_2 q_2 + b_3 q_3 - \nu t)} \qquad (11)$$

Here ν and the b_i are constants whose values we assume to be connected by the relation (10). The expression (11) then denotes a set of *monochromatic plane waves* spreading in the direction of the vector with the components b_i having the velocity $w = \frac{\nu}{b}$ and satisfying the frequency wave length condition (10).

The number of wave periods on a straight line of unit length crossing the parallel wave planes aslant is given by the component b_i taken for the direction considered. As usual, the wave is expressed by the use of an exponential expression with an imaginary exponent. This is a mathematical device used also for waves of other kinds, for instance, sound waves; it represents a method of abbreviating expressions written in terms of trigonometrical functions. By superposing several wave sets of the form (11) possessing certain properties of symmetry, the imaginary part of the wave expression is usually eliminated. What distinguishes the waves to be considered here from other waves is the fact that even in a superposition of sets like (11) the amplitude ψ is a complex number; thus the imaginary part of the resulting expression cannot be eliminated. The meaning of this feature, which is an essential part of wave mechanics, will be made clear in § 20.

The assumption (11) permits us to express the derivatives of ψ as follows:

$$\frac{1}{2\pi i} \frac{\partial \psi}{\partial q_k} = b_k \psi \qquad \frac{1}{(2\pi i)^2} \frac{\partial^2 \psi}{\partial q_k{}^2} = b_k{}^2 \psi \qquad -\frac{1}{2\pi i} \frac{\partial \psi}{\partial t} = \nu \psi \qquad (12)$$

These simple relations make it possible to transcribe the frequency wave-length equation (10) into a *differential equation for waves*. For this purpose we multiply every term in (10) by ψ, and then substitute the values (12). We thus obtain the relation:

$$-\frac{h}{2\pi i}\frac{\partial \psi}{\partial t} = \frac{h^2}{2m(2\pi i)^2}\left(\frac{\partial^2 \psi}{\partial q_1^2} + \frac{\partial^2 \psi}{\partial q_2^2} + \frac{\partial^2 \psi}{\partial q_3^2}\right) + U \cdot \psi \quad (13)$$

If, in particular, $U = 0$, we arrive at the equation

$$-\frac{h}{2\pi i}\frac{\partial \psi}{\partial t} = \frac{h^2}{2m(2\pi i)^2}\left(\frac{\partial^2 \psi}{\partial q_1^2} + \frac{\partial^2 \psi}{\partial q_2^2} + \frac{\partial^2 \psi}{\partial q_3^2}\right) \quad (14)$$

This is the *time-dependent Schrödinger equation for waves corresponding to free particles*.

Equation (13) is not essentially different from (14), since we assumed U to be constant. A difference results only with respect to the law of dispersion (10), relating the constants ν and b_i; if the potential field U is not zero, the waves possess a velocity w different from the velocity of the case $U = 0$.

Now let us turn to the more general case where the potential is not constant, but is a function $U(q_1,q_2,q_3)$ of the spatial coordinates alone. The derivation given does not cover this case, for the following reason. If we consider U in (10) as a function of the q_i, the b_i cannot be constants; but then the relations (12) do not hold. The method of derivation used here, therefore, does not tell us which kind of differential equation we must use for the case of a variable potential U. It is here that the method of derivation was to be replaced by "the darkness of instinctive guessing". Schrödinger saw that equation (13) can be extended, its form unchanged, to the case of a variable potential, and that thus the waves of the general case are controlled by the equation

$$-\frac{h}{2\pi i}\frac{\partial \psi}{\partial t} = \frac{h^2}{2m(2\pi i)^2}\left(\frac{\partial^2 \psi}{\partial q_1^2} + \frac{\partial^2 \psi}{\partial q_2^2} + \frac{\partial^2 \psi}{\partial q_3^2}\right) + U(q_1,q_2,q_3) \cdot \psi \quad (15)$$

The solutions of this equation, called the *time-dependent Schrödinger equation for waves corresponding to particles in a field of force*, do not have the simple form (11); this follows from the remarks just made. Schrödinger saw that, instead, solutions of a more complicated kind exist and that these solutions possess the mathematical properties required for a theory of the Bohr atom.

To show this, Schrödinger considered solutions of the form

$$\psi = \varphi(q_1,q_2,q_3)e^{-\frac{2\pi i}{h}Ht} \quad (16)$$

The solution (11) has this form when we put

$$\varphi(q_1,q_2,q_3) = \psi_0 e^{2\pi i(b_1q_1+b_2q_2+b_3q_3)} \quad (17)$$

since $\nu = \dfrac{H}{h}$. However, (16) is of a more general kind, since φ is not bound to

§13. SCHRÖDINGER'S EQUATION 71

the special form (17). Solutions of (15) in the general form (16) do exist, although, as we saw, there are no solutions satisfying (17). Let us assume that such a solution of the general form (16) is given. Carrying out the differentiation on the left hand side of (15) and canceling the exponential factor of (16), which does not depend on q, in all terms, we arrive at the equation

$$H \cdot \varphi = \frac{h^2}{2m(2\pi i)^2}\left(\frac{\partial^2 \varphi}{\partial q_1^2} + \frac{\partial^2 \varphi}{\partial q_2^2} + \frac{\partial^2 \varphi}{\partial q_3^2}\right) + U(q_1,q_2,q_3) \cdot \varphi \qquad (18)$$

This is the *time-independent Schrödinger equation for waves corresponding to particles in a field of force*. It establishes a particular rule for the spatial part φ of the ψ-function. Schrödinger recognized that this equation, if some requirements of regularity are added, has, in general, solutions only for discrete values of the constant H and therefore determines discrete energy values which correspond to Bohr's energy levels of the atom.

We certainly do not disparage Schrödinger's work when we do not consider the derivation given, or the more complicated derivations originally given by Schrödinger, as a proof of the validity of the resulting wave equation. Such a derivation—and Schrödinger never meant anything else—can be used to make the wave equation *plausible*, and therefore represents an excellent guide within the context of discovery. In our exposition it is the extension of (13) to the case (15) of a variable potential $U(q_1,q_2,q_3)$ which cannot be justified by deductive reasoning. But even the derivation of the wave equation (13) for the particle in a constant potential field includes assumptions which by no means can be taken for granted. Thus we have no a priori evidence for the assumption that equation (9), after the introduction of the Planck-de Broglie relations, will hold strictly. Judging without further knowledge of quantum mechanics, one might well suppose that waves are controlled by a more complicated relation than (10), and that the introduction of particle concepts like position and momentum is connected with simplifications which confer to (9) an approximative character. A proof that this is not the case can be given only by showing that the consequences of this equation are verified by observations.

It is only after this proof has been given that we can consider as valid the principles used in the derivation of the equation, thus evaluating the derivation in the reverse sense. We must limit, however, this evaluation to the wave equation for free particles, since only the inferences used in the derivation of this equation can be reversed. We then can state the bearing of the derivation given as follows: A wave interpretation of free particles, using waves which satisfy Schrödinger's equation, can always be translated through the Planck-de Broglie principle into a corpuscle interpretation satisfying the energy-momentum relation. In other words, *Schrödinger's wave equation guarantees the strict duality of wave and corpuscle interpretation for free particles*.

Whereas the principle of duality has turned out to be a good guide for the establishment of the wave equation, it appears advisable not to use this principle for the presentation of the system of rules which is now generally used for the solution of quantum mechanical problems. First, it is clear from the derivation given that the duality is limited to free particles, i.e., particles whose energy is composed of kinetic energy and a constant amount of potential energy. Particles in a field of force do not satisfy the energy-momentum relation (9), as will be discussed in § 33. Second, it has turned out that the principle is of no help so far as the meaning of the function ψ is concerned; the rules translating this function into probabilities cannot be derived from the principle of duality. Third, it seems appropriate to begin the presentation, not with the general equation (15), but with the special equation (18), and to introduce this relation in a form not restricted to the energy H, but holding for all kinds of physical entities. The justification of the latter extension is given, as before, by its success.

We therefore now shall turn to an exposition of the system of quantum mechanical rules which is not governed by the desire to make these rules plausible, nor by the intention to show the origin of these rules in physical conceptions grown from older quantum physics. This exposition will not refer to the derivation of the wave equation given in this section. We leave it to the reader to recognize in the rules to be presented the path of the derivation given.

§ 14. Operators, Eigen-Functions, and Eigen-Values of Physical Entities

The *rules of transcription* which coordinate quantum mechanical laws to classical laws are stated by means of *operators*. The term "operator" is used here in the same sense as it has been used in our explanation of transformations (§ 10, § 12), namely, as meaning a rule which coordinates to a given function $\psi(q)$ another function $\chi(q)$. The operators used in quantum mechanics are generally constructed out of two elementary operators: the differential quotient $\frac{\partial}{\partial q}$, and multiplication by q. These expressions are operators because they coordinate to a given function $\psi(q)$, respectively, the functions $\frac{\partial \psi}{\partial q}$ and $q \cdot \psi(q)$.

The construction of more complicated operators is established in the following way.

We assume that a mechanical problem is given, in the classical form, in the canonical parameters $q_1 \ldots q_n$, $p_1 \ldots p_n$. A physical entity u is determined, classically speaking, if it is given as a function of these parameters:

$$u = u(q_1 \ldots q_n, p_1 \ldots p_n) \tag{1}$$

We may assume that u is a polynomial in the p_i. We now coordinate

to q_i the operator: multiplication by q_i

§ 14. OPERATORS

to $f(q_i)$ the operator: multiplication by $f(q_i)$

to p_i the operator: $\dfrac{h}{2\pi i}\dfrac{\partial}{\partial q_i}$ (2)

to p_i^2 the operator: $\dfrac{h^2}{(2\pi i)^2}\dfrac{\partial^2}{\partial q_i^2}$

where $f(q_i)$ means any function of q_i, and operators of higher powers of p_i are defined in continuation of the rule given for p_i^2. Putting these operators in the place of the coordinates q_i and p_i within the function $u(q_1 \ldots q_n, p_1 \ldots p_n)$, we obtain a compound operator u_{op}. We have thus coordinated to the entity u, instead of the function $u(q_1 \ldots q_n, p_1 \ldots p_n)$ of (1), an operator u_{op}.

For instance, let u be the energy H, and assume that the function $H(q_1 \ldots q_n, p_1 \ldots p_n)$ has the form (9), § 13. Our rule (2) then coordinates to (9), § 13, the operator

$$H_{op} = \frac{1}{2m}\left(\frac{h}{2\pi i}\right)^2\left(\frac{\partial^2}{\partial q_1^2} + \frac{\partial^2}{\partial q_2^2} + \frac{\partial^2}{\partial q_3^2}\right) + U(q_1, q_2, q_3) \quad (3)$$

where $U(q_1, q_2, q_3)$ is now an operator meaning multiplication with this function.

We say that the entity u is represented by the operator u_{op} within the physical context stated by the equation (1). It is important to realize that we cannot speak without further qualification of an operator coordinated to an entity; the operator is determined only if the entity occurs within a given context, i.e., if the entity is defined as a function of the p_i and q_i. This context can only be stated classically when we formulate it in ordinary language. If this reference to classical mechanics is to be avoided, we must consider the operator as the definition of the physical context.

The operator is a mathematical instrument which we use to coordinate to the entity it represents: *eigen-functions* and *eigen-values*. This is done by means of a differential equation constructed in terms of the operator, called the *first Schrödinger equation*. It has the same form for all physical entities; the differences between the various entities find their expression only in the nature of the operator. We therefore also speak of the eigen-functions and eigen-values of the operator.

The first Schrödinger equation, which is also called the *time-independent* Schrödinger equation, has always the following form:

$$u_{op}\varphi(q) = u \cdot \varphi(q) \quad (4)$$

Since this equation is required to hold in combination with certain requirements of regularity, such as the requirement that the function φ be finite for all values of its argument, it has, in general, solutions φ only for certain discrete values u_i of the constant u; these values u_i constitute the eigen-values of the entity u. We speak in such a case of a *discrete spectrum* of *eigen-values*. To each value u_i belongs a function $\varphi_i(q)$ as a solution of the equation; these functions $\varphi_i(q)$ constitute the *eigen-functions*. We can write (4) for this case in the form

$$u_{op}\varphi_i(q) = u_i \cdot \varphi_i(q) \tag{5}$$

In other cases we have solutions for a continuous manifold of values u; we then say that we have a *continuous spectrum* of eigen-values. The solutions $\varphi(q)$ contain, of course, the constant u; they therefore have the form $\varphi(u,q)$ and form a continuous set of orthogonal functions. The Schrödinger equation (4) then assumes the form

$$u_{op}\varphi(u,q) = u \cdot \varphi(u,q) \tag{6}$$

We must realize that here the entity u has the character of a constant with respect to the operator u_{op}, since this operator, according to (2), contains only operations in the variable q. This constant u takes the place of the subscript i in (5) and classifies the solutions φ of (4).

It may happen that, for one eigen-value u_i, (5) furnishes, not one, but n solutions $\varphi_i(q)$. This is called a case of *degeneracy of the degree n*.

In order to show the method at work we continue our example. If we insert the operator (3) in the general equation (4), we obtain the differential equation (18), § 13. In combination with requirements of regularity the equation has solutions only for a set of discrete values H_i, the eigen-values of the energy; the coordinated solutions $\varphi(q_1,q_2,q_3)$ constitute the eigen-functions of the energy. Examples of the latter functions, constructed for special forms of the potential $U(q_1,q_2,q_3)$, are given in the textbooks of quantum mechanics.

By means of the mathematical technique so far developed we can coordinate to every physical entity within a given context a set of eigen-values and eigen-functions. The mathematical importance of these concepts consists in the fact that *the eigen-functions defined by the first Schrödinger equation constitute an orthogonal set, and that the eigen-values are real numbers*.

The proof of this theorem requires some additional remarks about operators. Using the geometrical mode of speech introduced in § 9, we can regard an operator as a transformation in the active interpretation of this term; and we can therefore say that the expression $u_{op}\varphi$ denotes a vector coordinated to the vector φ by the deformation of the space. The Schrödinger equation (4), in this geometrical interpretation, states the condition that the series of deformations of the space expressed by the operator u_{op} is of such a kind that ultimately the vector φ is transformed into a simple multiple $u \cdot \varphi$ of itself.

The operators of quantum mechanics are all of a special type: They are *linear* and *Hermitean*. An operator u_{op} is called *linear* if for any vectors φ and χ it satisfies the relations:

$$\left. \begin{aligned} u_{op}(\varphi + \chi) &= u_{op}\varphi + u_{op}\chi \\ u_{op}(\kappa \cdot \varphi) &= \kappa \cdot u_{op}\varphi \end{aligned} \right\} \tag{7}$$

§ 14. OPERATORS   75

where κ is any complex number (not a vector). An operator is called *Hermitean* if for any two vectors φ and χ the relation holds:[1]

$$(u_{op}\varphi, \chi) = (\varphi, u_{op}\chi) \tag{8}$$

where the parentheses denote the inner product, as above. It can be shown that the operators defined in (2) are linear and Hermitean. If the latter condition is not automatically satisfied by the construction of further operators through the rules (2), the form of the function (1) is interpreted in such a way that the operator is made Hermitean. A corresponding qualification is to be added to the rules determining the construction of operators. It can be shown, furthermore, that the energy operator (3) is Hermitean.

We now can prove that the solutions of (4) constitute an orthogonal set, and that the eigen-values are real numbers. To prove the latter condition, let us start from the relation

$$(u_{op}\varphi, \varphi) - (\varphi, u_{op}\varphi) = 0 \tag{9}$$

which follows from (8) if we take the special case $\chi = \varphi$. With (4) this can be transformed into

$$(u\varphi, \varphi) - (\varphi, u\varphi) = 0 \tag{10}$$

Since u is a number (not a vector), we can apply (21), § 10, and thus derive:

$$u \cdot (\varphi,\varphi) - u^* \cdot (\varphi,\varphi) = 0$$

$$u - u^* = 0 \tag{11}$$

The division by (φ,φ) is permissible since $(\varphi,\varphi) = 0$ holds only if the function φ vanishes, a case which need not be considered. The result means that u is a real number. To prove the orthogonality of the set, let us assume two functions φ_1 and φ_2 which constitute solutions of (4); we then have, because of (8) and (4):

$$0 = (u_{op}\varphi_1, \varphi_2) - (\varphi_1, u_{op}\varphi_2) = (u_1\varphi_1, \varphi_2) - (\varphi_1, u_2\varphi_2)$$
$$= u_1(\varphi_1,\varphi_2) - u_2^*(\varphi_1,\varphi_2) = (u_1 - u_2) \cdot (\varphi_1,\varphi_2) \tag{12}$$

Since u_1 and u_2 are assumed to be different from each other, the inner product (φ_1,φ_2) must vanish. This expresses the orthogonality of these functions. The condition $(\varphi,\varphi) = 1$ can be easily satisfied by multiplying the function by a suitable constant of normalization, since the definition of vectors requires that (φ,φ) must be finite.

The given proof cannot be applied to the case of continuous eigen-values u for the following reasons. Eigen-functions $\varphi(u,q)$ do not possess a finite $\int |\varphi(u,q)|^2 dq$, as is visible also from (16) and (17), § 9, the $\delta(x,z)$-symbol being infinite for $x = z$. Such functions, therefore, do not constitute vectors in the

---

[1] The name "Hermitean" is derived from the name of the French mathematician Hermite. It can be shown that a Hermitean operator must be linear, whereas, of course, a linear operator need not be Hermitean.

Hilbert space, and thus cannot be substituted for φ and χ in (8). The result $(\varphi_1,\varphi_2) = 0$ is therefore not derivable for them, in correspondence with the fact that these functions are not reflexive (cf. § 9). Functions of this kind are the Fourier-functions $const \cdot e^{\frac{2\pi i}{h} pq}$ which represent the eigen-functions of the momentum p. The eigen-value problem of such functions has been discussed by J. v. Neumann;[2] it can be shown that also the continuous eigen-spectrum of the Schrödinger equation constitutes an orthogonal set.

§ 15. The Commutation Rule

We now must consider a further property of operators. If an operator u_{op} is applied to a function φ, it produces a new function; to this function we can apply another operator v_{op} and thus produce a third function. This iterated application of operators is expressed in the form

$$v_{op} u_{op} \varphi \tag{1}$$

The combination $v_{op} u_{op}$ can be considered as a new operator which is called the product of the two operators. Here the word "product" is, of course, used in a generalized sense, similar to the term "relational product" of logistic. It is obvious that, in general, the application of the two operators in the reversed order will not lead to the same function; i.e., in general, the expression

$$v_{op} u_{op} \varphi - u_{op} v_{op} \varphi = [v_{op} u_{op} - u_{op} v_{op}] \varphi \tag{2}$$

will not be equal to 0. The multiplication of operators, therefore, is, in general, not commutative. There is, of course, nothing paradoxical in such a statement if we realize the meaning of the word "product", or "multiplication", used here.

Usually the expression (2) will be different from zero for most functions φ, and equal to zero only for some special functions φ. When, for two particular operators, the expression (2) vanishes for *all* functions φ, this will represent a specific property of these operators; they then are called *commutative operators*. Similarly, the entities to which these operators belong are called *commutative entities*. Regarding the expression in the brackets in (2) as one operator, which we call the *commutator*, we can express this property symbolically by saying that the commutator of the two operators vanishes, or satisfies the equality

$$v_{op} u_{op} - u_{op} v_{op} = 0 \tag{3}$$

On the other hand, operators for which the expression (2) does not vanish identically in the φ, or, what is the same, for which there are functions φ such that (2) is not $= 0$, are called *noncommutative*. We characterize such operators by saying that their commutator does not vanish, or that for it the inequality

$$v_{op} u_{op} - u_{op} v_{op} \neq 0 \tag{4}$$

holds. The corresponding entities are called *noncommutative entities*.

[2] *Mathematische Grundlagen der Quantenmechanik* (Berlin, 1932), II: 6–9.

§ 15. THE COMMUTATION RULE

It can be shown that commutative operators or entities have the same system of eigen-functions, although, of course, they have different eigen-values. This case holds for entities which are functions of each other. For instance, the entities u and u^2 have the same eigen-functions, but different eigen-values. Noncommutative operators or entities, on the contrary, have different systems of eigen-functions.

The proof follows when we apply the operator u_{op} to both sides of (4), § 14:

$$v_{op} u_{op} \varphi = v_{op} u \varphi$$

When the operators are commutative we can reverse their order on the left hand side; on the right hand side, using (7), § 14, we can put u before the operator. We thus obtain

$$u_{op} v_{op} \varphi = u \cdot v_{op} \varphi$$

When we regard here the expression $v_{op}\varphi$ as a unit, this equation states that the function $v_{op}\varphi$ satisfies (4), § 14. Disregarding the case of degeneracy, we therefore can infer that $v_{op}\varphi$ must be identical with φ apart from a constant factor (since (4), § 14, determines φ only to a constant factor). Calling this constant v, we thus have

$$v_{op}\varphi = v\varphi$$

But this relation has the form (4), § 14, for the operator v_{op}, and therefore states that φ is also an eigen-function of v_{op}. It is easily seen that this proof can be reversed; i.e., that starting with the assumption of identical eigen-functions, we can show that the operators are commutative. It follows that noncommutative operators have different eigen-functions. Let us add the remark that the proof can be extended to include the case of degeneracy.

The distinction of these two kinds of operators finds an important application with respect to the elementary operators p_{op} and q_{op} defined in (2), § 14. Applying these two operators (constructed for the same subscript i, i.e., for coordinated values p_i and q_i) to a function φ we have

$$p_{op} q_{op} \varphi - q_{op} p_{op} \varphi = \frac{h}{2\pi i} \left\{ \frac{\partial}{\partial q} (q \cdot \varphi) - q \cdot \frac{\partial \varphi}{\partial q} \right\} = \frac{h}{2\pi i} \varphi \qquad (5)$$

We write this in the form

$$p_{op} q_{op} - q_{op} p_{op} = \frac{h}{2\pi i} \qquad (6)$$

This equation is called the *commutation rule*. It states the noncommutativity of the momentum-operator and the position-operator. In contradistinction to this case it can easily be shown that two operators of the same kind, such as the operators q_1 and q_2, or the operators p_1 and p_2, are commutative. The two noncommutative parameters p_i and q_i are also called *canonically conjugated parameters* or *complementary parameters*. Two parameters p_i and q_k, which are not canonically conjugated (i.e., which have different subscripts), are commutative.

The p and q are the fundamental noncommutative operators. There are also other operators which are noncommutative with each other. An entity which is a function of the p alone is commutative with the p, but not with the q; and vice versa a function of the q alone is commutative with the q, but not with the p. Functions of both q and p, however, can be noncommutative with both q and p; an example is given by the angular momentum.[1]

§ 16. Operator Matrices

Before we turn to the application of the results so far obtained we shall first explain a second method of determining eigen-values and eigen-functions.

The geometrical interpretation of operators suggests the supposition that linear operators can be replaced by matrix transformations such as explained in § 9. This supposition can be shown to be true for the discrete case. An expression like $u_{op}\varphi$ then is replaced by a summation, or integration, in terms of a matrix.

For this purpose we coordinate to the operator u_{op} an operator matrix. We call this matrix τ_{ik}. The components of τ_{ik} are defined by the application of u_{op} to the unit vector $\epsilon_{(i)}$ (cf. § 10). We define

$$\tau_{ki} = [u_{op}\epsilon_{(i)}]_k \qquad (1)$$

It then can be shown that the application of the operator u_{op} to a discrete function χ_i is equivalent to a summation in the following sense

$$[u_{op}\chi]_k = \sum_i \chi_i \tau_{ki} \qquad (2)$$

The left hand side means the k-th component of the function resulting from the application of u_{op} to χ.

We can show that condition (8), § 14, of the Hermitean character of operators is equivalent to the condition that the operator matrix τ_{ik} satisfies the condition

$$\tau_{ki} = \tau_{ik}^* \qquad (3)$$

Writing (8), § 14, as a summation according to (18), § 10, we have with (2)

$$0 = (u_{op}\varphi, \chi) - (\varphi, u_{op}\chi) = \sum_k [u_{op}\varphi]_k \cdot \chi_k^* - \sum_i \varphi_i [u_{op}\chi]_i^*$$
$$= \sum_k \sum_i \tau_{ki}\varphi_i\chi_k^* - \tau_{ik}^*\varphi_i\chi_k^* = \sum_k \sum_i (\tau_{ki} - \tau_{ik}^*)\varphi_i\chi_k^* \qquad (4)$$

If this is to hold for any two vectors φ and χ, the expression in the parentheses in the last term must vanish. We therefore have $\tau_{ki} = \tau_{ik}^*$.

Whereas in the discrete case we can always coordinate an operator matrix

---
[1] Cf. H. A. Kramers, *Grundlagen der Quantentheorie* (Leipzig, 1938), p. 166.

§16. OPERATOR MATRICES

to a linear operator, this is not always possible for the continuous case. If there is such a matrix $\tau(q',q)$ the application of the operator is equivalent to an integration in the following sense:

$$[u_{op}\chi]_{q'} = \int \chi(q)\tau(q',q)dq \tag{5}$$

The left hand side means the value of the resulting function at the place q'. It can be shown, as before, that Hermitean character of the operator, defined by (10), § 13, is equivalent to the condition:

$$\tau(q',q) = \tau^*(q,q') \tag{6}$$

The advantage of condition (8), § 14, over (3) and (6) consists in the fact that (8), § 14, can always be applied, even if no operator matrices exist.[1] For operators of the heterogeneous type, i.e., operators which coordinate a discrete function to a continuous function, or vice versa, Hermitean character cannot be defined.

In cases where a continuous operator matrix does not exist, we can at least introduce matrices which approximately possess the required properties. This approximation is reached by means of the *Dirac functions* (cf. § 9). It turns out that both the elementary operators q and $\dfrac{\partial}{\partial q}$ require a treatment of this kind and have Dirac functions as their matrix functions. The matrix function of the operator q is the function $q \cdot \delta(q',q)$; that of the operator $\dfrac{\partial}{\partial q}$ is the first derivative $\delta'(q',q)$.

In order to construct the operator matrix of a physical entity, it is not necessary to construct first the Schrödinger operator by means of the rules (2), § 14, and then to use (1). Starting from the classical function (1), § 14, one can directly proceed to the construction of the matrix τ_{ik} of the entity u when the matrices coordinated to the entities p and q are given; the matrix τ_{ik} then is built up of elements written as functions of the matrix elements of p and q. We shall not give here the rules necessary for this construction of operator matrices, but refer the reader to the textbooks of quantum mechanics. A further rule in this *matrix mechanics* is the commutation rule (6), § 15; since all other operator matrices are functions of the matrices p and q, this basic property of the operators p and q finds an expression in the structure of all other matrices.

---

[1] It is possible to construct matrices which are at the same time Hermitean and unitary. For such matrices we have with (2), § 11, and (15), § 11, and (3) and (6):

$$\varphi_{ik} = \varphi_{ik}^{-1} \qquad \varphi(q',q) = \varphi^{-1}(q',q)$$

The matrices of quantum mechanics, however, usually satisfy only one of the two conditions. Matrices, or operators, coordinated to physical entities, are Hermitean; matrices used as eigen-functions, or transformations, are unitary.

Once the operator matrices of physical entities have been constructed, they can be used to determine eigen-values and eigen-functions of the entities. The method used for this purpose is once more independent of the Schrödinger method and thus does not make use of the Schrödinger equation; instead, it is based on the following postulate.

Matrix postulate: Construct a unitary matrix φ_{ik} which transforms τ_{ik} into a diagonal matrix $u_i \cdot \delta_{ik}$, i.e., a matrix whose values are different from zero only in the diagonal lines $i = k$; the values u_i of this diagonal matrix are the eigen-values of u, and the matrix φ_{ik} represents the eigen-functions of u.

The eigen-values so defined are real numbers because of the Hermitean character of τ_{ik} and the unitary character of φ_{ik}; the latter property guarantees, at the same time, that the φ_{ik} constitute a complete orthogonal system. The fundamental role which the commutation rule plays in this method is clear from the following result: Although the elements τ_{ik} depend on the elements of the matrices p and q, the eigen-values of τ_{ik}, i.e., the diagonal matrix into which τ_{ik} is transformed, will not depend on the elements of p and q if only the latter matrices satisfy the commutation rule. In other words: The latter rule introduces a sufficient number of relations between the matrix elements of p and q in order to enable us to eliminate these matrix elements from the eigen-values of τ_{ik}. It is the commutation rule, therefore, which makes the eigen-values of physical entities definite.

For continuous operator matrices $\tau(u,q)$ a similar formulation can be given.

It was this matrix mechanics which was introduced by Heisenberg, Born, and Jordan,[2] and separately by Dirac,[3] independently of the wave theory of L. de Broglie and Schrödinger. Schrödinger, after having constructed the differential equation (4), § 14, in the pursuit of wave conceptions, showed in a consecutive paper[4] that this differential equation is mathematically equivalent to the postulate of matrix mechanics, since the eigen-values and eigen-functions of the operator, defined by (4), § 14, are identical with those defined by the matrix postulate. Mathematically speaking, the Schrödinger method has the advantage that it is easier to solve the differential equation (4), § 14, than to find a unitary matrix φ_{ik} of the required properties; and actually the construction of such unitary matrices is usually given by the use of the Schrödinger equation. This is the reason that now the matrix form of quantum mechanics is usually replaced by the operator form, i.e., a form in which the equivalence (2), or (5), between operator and matrix summation is not used. The eigen-values and eigen-functions of operators are then defined by the Schrödinger equation (4), § 14, which makes the use of the matrix postulate unnecessary.

[2] *Zeitschr. f. Phys.*, 35 (Berlin, 1926), p. 557.
[3] *Proc. Roy. Soc. London*, (A), Vol. 109 (1925), p. 642.
[4] *Ann. d. Phys.* (4), Vol. 79 (Leipzig, 1926), p. 734.

§ 17. Determination of the Probability Distributions

Eigen-values and eigen-functions characterize physical entities within a given physical context. We must, however, distinguish the *physical context*, as the totality of structural relations, from the *physical situation*, which includes, in addition to the structure, a determination of the probability distributions of the entities involved.

Classically speaking, this distinction refers to the *functional relations* of a physical problem on the one side, and the *numerical values* of the entities on the other side. The totality of functional relations determines the context; but the situation is known only if, in addition, the numerical values of the entities are given. In quantum mechanics the functional relations are replaced by the *structure of the operators* which, in combination with the first Schrödinger equation, determine the eigen-values and eigen-functions. The indication of numerical values, on the other hand, is replaced by the indication of the *probability distributions* of the entities. This corresponds to the transition from causal laws to probability laws, which expresses the outstanding feature of quantum mechanics.

The characterization of the physical situation therefore requires a mathematical instrument which, in addition to the structure expressed in terms of eigen-values and eigen-functions, indicates the probability distributions. This instrument is the state function ψ. Postponing the question how this function is to be ascertained, we shall now explain the way in which it determines the probability distributions.

The state function ψ is a function of the coordinates q; in addition, it will, in general, depend on the time t. It therefore has the form $\psi(q_1 \ldots q_n, t)$, or $\psi(q,t)$. The second term may be considered as an abbreviation, in which we write, as before, q for $q_1 \ldots q_n$. We shall turn to the analysis of the time dependence of ψ later (§ 18); for the present we shall therefore omit the argument t, and simply write $\psi(q)$. We may understand this as meaning that we choose a special value of t, and consider the relations then ensuing between the so-specialized function ψ and the probability distributions; the same relations will hold for every value of t.

A general requirement concerning ψ is that this function is *quadratically integrable and normalized*, i.e., that it satisfies the relation

$$\int |\psi(q)|^2 dq = 1 \tag{1}$$

According to our statement concerning time dependence, this condition is supposed to hold for every value t. The meaning of this condition, which is closely connected with the probability interpretation, will turn out presently.

Before we determine the probabilities of observable values we must first state which are the *possible* values. This is necessary because in quantum mechanics, in contradistinction to classical physics, not all numerical values are, in general, physically possible values. The desired statement is given by the following rule:

I. *Rule of the eigen-values. The eigen-values constitute the possible values of the entity within the given context.*

An example of a discrete spectrum of eigen-values is given by the energy levels in an atom, which are determined by the Schrödinger equation (5), § 14, written for the operator H_{op}. The sensational effect of Schrödinger's discovery was due to the fact that with his equation Bohr's discrete stationary states of the atom were shown to be interpretable as an eigen-value problem. Only if the spectrum of eigen-values is continuous and unlimited, does the list of possible values correspond to the classical case. Of such kind is the eigen-value spectrum of the position coordinates of a free particle. In most cases the eigen-value spectrum is partly discrete, partly continuous. In the hydrogen atom, for instance, discrete spectra are produced by transitions from one orbit to another; the continuous spectrum corresponds to the capturing of free electrons, or to the reversed process, i.e., ionization.

Whereas the eigen-values have an immediate meaning with respect to observations, the eigen-functions have not; they constitute only a mathematical instrument necessary for the derivation of the probability distributions. These distributions, to which we now must turn, are determined by the following rule, which has been introduced by Born, and which we have already mentioned in a special form in § 2:

II. *Rule of spectral decomposition. Expand the function $\psi(q)$ into the eigen-functions of the considered entity; the squares of the expansion coefficients σ_i, or $\sigma(u)$, then determine the probability that an eigen-value u_i, or u, is to be observed.*

This rule is expressed by the following formulae:

Discrete case:
$$\psi(q) = \sum_i \sigma_i \varphi_i(q) \qquad (2)$$

$$P(s, u_i) = |\sigma_i|^2 \qquad (3)$$

Continuous case:
$$\psi(q) = \int \sigma(u) \varphi(u, q) du \qquad (4)$$

$$P(s, u) = |\sigma(u)|^2 \qquad (5)$$

The symbol P stands for "probability". Since these probabilities are relative to the physical situation s characterized by the function ψ, we have put the symbol "s" into the first term of the probability expression. Instead of using

§ 17. DETERMINATION OF DISTRIBUTIONS

the logical symbol P, we can also express a probability by means of the mathematical symbol "d", which denotes a probability function. We then have, instead of (3) and (5), the relation:

$$d(u_i) = |\sigma_i|^2 \qquad (6)$$

$$d(u) = |\sigma(u)|^2 \qquad (7)$$

The function d depends on the situation s; if necessary this can be expressed by a subscript, in the form $d_s(u)$. (3) and (6) are probabilities; (5) and (7) are probability densities, i.e., they are turned into probabilities by integration over u, between any two limits u_1 and u_2.

From the exposition of the theory of expansions given in § 9, it is clear that the probabilities (3) and (5) do not depend upon q, since σ does not. Because of (1) and (3), § 9, and (15), § 9, the probabilities satisfy the condition

$$\sum_i |\sigma_i|^2 = 1 \qquad (8)$$

$$\int |\sigma(u)| du = 1 \qquad (9)$$

This condition is necessary because one of the eigen-values must be realized.

For the case that an eigen-value is degenerate (cf. p. 74) the expansion (2) assumes a somewhat more complicated form. Let $\varphi_{i_s}(q)$ be the n eigen-functions belonging to the eigen-value u_i; then the term $\sigma_i \varphi_i(q)$ of the expansion (2) is replaced by a term

$$\sigma_i \sum_{s=1}^{n} \sigma_{i_s} \varphi_{i_s}(q) \qquad (10)$$

The σ_{i_s} satisfy the condition

$$\sum_{s=1}^{n} |\sigma_{i_s}|^2 = 1 \qquad (11)$$

The expression $\sum_{s=1}^{n} \sigma_{i_s} \varphi_{i_s}(q)$ can be formally treated like a term $\varphi_i(q)$. We therefore shall not introduce a special indication for the possibility of degeneracy in the following formulae. An illustration of degeneracy is given by a Bohr atom in which the same total energy H_i can result with various arrangements of the electrons in different orbits. Each such arrangement constitutes a determinate physical state and is characterized by one of the eigen-functions $\varphi_{i_s}(q)$. To one eigen-value H_i there correspond here different states $\varphi_{i_s}(q)$; the probabilities of these states are given by $|\sigma_i \sigma_{i_s}|^2$.

For the continuous case (5), (7), and (9), we must add the qualification that, if u is not suitably chosen, a density function $r(u)$ according to (22), § 9, will appear; we then have

$$P(s,u) = d(u) = |\sigma(u)|^2 r(u) \qquad (12)$$

$$\int |\sigma(u)|^2 r(u) du = 1 \qquad (13)$$

84   PART II. MATHEMATICAL OUTLINES

The density function $r(u)$ can always be eliminated if we introduce, instead of u, a suitable function u' of u.

We now must add a rule determining the probability distribution of the entities used as arguments of the ψ-function. We have here the rule mentioned already in § 2:

III. *Rule of the squared ψ-function.* The probability of a value q (or of a set of values $q_1 \ldots q_n$) which is an argument of the ψ-function is given by the expression

$$P(s,q) = |\psi(q)|^2 \qquad (14)$$

This rule makes clear the meaning of the normalization (1). The integral of $|\psi|^2$ taken over all values q means the probability of observing any value q at all, and this probability must be $= 1$.[1]

Rules I–III hold for every quantum mechanical system, and for every time point t. These rules connect the abstract mechanism of formulae with observational entities. It is this fact from which they derive their significance. Since the probabilities are derived always as the squares of complex functions, these latter functions are sometimes called *probability amplitudes*.

It can be shown that rule III can be formally included in rule II. The eigenfunctions $\varphi(u,q)$ of the entities u considered contain the entities u and the coordinates q as their arguments; they therefore represent relations between u and q. It is formally possible also to introduce eigen-functions of the q; these functions then represent relations in which the entity q occupies both places of the function. We express these functions by $\varphi(q',q)$. By putting the operator q_{op} in the place of u_{op} we can even use the Schrödinger equation (5), § 14, to define these functions; the equation then furnishes the Dirac functions $\delta(q',q)$ as representing the $\varphi(q',q)$. Expanding $\psi(q)$ into these eigen-functions we have the trivial relation (cf. (14), § 9):

$$\psi(q) = \int \psi(q')\delta(q',q)dq' \qquad (15)$$

We see that rule II, applied to this expansion, immediately furnishes (12), i.e., rule III.

Rule II in combination with the expansion of $\psi(q)$ in terms of the φ and the Schrödinger equation (5), § 14, permits us to derive a simple relation for averages of physical entities, which is frequently used in quantum mechanics. According to the rules of probability, the average of an entity u with the distribution $d(u)$ is given by the expression

$$Av(u) = \int u\, d(u) du \qquad (16)$$

---

[1] For certain physical problems condition (1) is abandoned, and a nonquadratically integrable ψ-function is used. This means that $|\psi|^2$ is interpreted, not as a probability, but as an instrument of determining ratios of probabilities.

§ 18. TIME DEPENDENCE

With (7), (4), and the complex conjugate relation of (11), § 9, we have

$$Av(u) = \int u\, d(u)du = \int u|\sigma(u)|^2 du = \int \sigma(u)u\, \sigma^*(u)du \tag{17}$$

$$= \iint \sigma(u)u\psi^*(q)\varphi(u,q)du\, dq \tag{18}$$

when we change the order of the integrations. Now we have with (4), and because of the linearity of u_{op} (note that $\sigma(u)$ has the character of a constant with respect to u_{op}):

$$u_{op}\psi(q) = u_{op}\int \sigma(u)\varphi(u,q)du = \int u_{op}\sigma(u)\varphi(u,q)du$$

$$= \int \sigma(u)u_{op}\varphi(u,q)du = \int \sigma(u)u\, \varphi(u,q)du \tag{19}$$

For the last step we have used the Schrödinger equation in the form (6), § 14. With (19) we can write (18) in the form

$$Av(u) = \int \psi^*(q)u_{op}\psi(q)dq \tag{20}$$

Since the average of an entity is sometimes called *expectation value*, this formula is often called *expectation formula*. Comparing (20) with the last expression in (17) we see that, formally speaking, the introduction of $\psi(q)$ in the place of $\sigma(u)$ is accompanied by the replacement of u by u_{op}.

§ 18. Time Dependence of the Ψ-Function

We now turn to considerations which concern the changes of the physical situation within the course of time, and which therefore deal with the dependence of ψ on t. Just as the numerical values of variables may change in classical physics, such as the position of a mass point in motion, or the angular deviation of a pendulum, the probability distributions of the entities can change in quantum mechanics.

It is here that we come to use the general Schrödinger equation introduced in (15), § 13, since this equation controls the time dependence of ψ. Whereas (15), § 13, is given for a special form of the energy, we shall now present the general form of this equation. Let us state here, as before, that the justification of this equation is given only by its empirical success. The *second Schrödinger equation*, or *time-dependent Schrödinger equation*, has the form

$$H_{op}\psi(q,t) = \frac{ih}{2\pi}\frac{\partial}{\partial t}\psi(q,t) \tag{1}$$

H_{op} means the operator of the energy. It is important that, whereas the first Schrödinger equation can be equally applied to operators belonging to all kinds of physical entities, the second Schrödinger equation is constructed only in terms of the energy operator.[1]

For an illustration of (1) we may use the special form (3), § 14, of the Hamiltonian operator H_{op}. (1) then assumes the form (15), § 13.

The form of the general equation (1) makes it necessary that the function ψ practically always depends on time. For a time-independent ψ the right hand side of (1) vanishes. Now the energy operator H_{op}, in all practical cases, is so constructed that the left hand side of (1) does not vanish.[2] Hence ψ must depend on time.

The significance of equation (1) consists in the fact that it determines the form of $\psi(q,t)$ for every time t, if the form of ψ at the time t_0, i.e., the function $\psi(q,t_0)$, is given. This property follows, because (1) is a differential equation of the first order in $\dfrac{\partial}{\partial t}$. The second Schrödinger equation therefore formulates the law which controls the change of the state function ψ with time.

We now must consider the consequences resulting from the time dependence of ψ in an expansion of the form (2), § 17, or (4), § 17. If ψ depends on t, we must introduce the argument t also on the right hand side of the expansion. This can be done in two ways: either the eigen-functions φ (case 1) or the coefficients σ (case 2) can be considered as depending on t. Both cases occur in quantum mechanics. They may also be combined such that both φ and σ depend on t (case 3).

The simplest case of time dependence is given by a ψ-function of the special form (16), § 13, i.e., by a function

$$\psi_k(q,t) = \varphi_k(q) \cdot e^{-\frac{2\pi i}{h} H_k t} \qquad (2)$$

which splits into a factor depending on q alone, and an exponential factor containing t and the eigen-value H_k of the energy, but not q. The inferences leading from the time-dependent equation to the time-independent one, given in §13 for a special form of the energy, can be repeated for the general form (1). Introducing the solution (2) into (1) we obtain

$$H_{op}\psi_k(q,t) = H_k\psi_k(q,t) \qquad (3)$$

This is the first Schrödinger equation (4), § 14, written for the operator H_{op}. We now can cancel the exponential factor on each side; on the left side this is possible because the exponential factor, which does not depend on q, has the character of a constant with respect to the linear operator H_{op}. We thus obtain
$$H_{op}\varphi_k(q) = H_k\varphi_k(q) \qquad (4)$$

---

[1] Originally Schrödinger had developed also his first equation only in terms of the energy operator. The extension to other operators was introduced later.
[2] If, in (9), § 13, the potential U is $=0$, the left hand side of (1) vanishes for a function ψ which is linear in q. But such a function cannot satisfy the condition of normalization (1), § 17. We therefore must have a time-dependent ψ also in this case.

§18. TIME DEPENDENCE

We see that, as before, the time-independent factor $\varphi_k(q)$ is determined by the first Schrödinger equation. If $\varphi_k(q)$ is chosen in correspondence with this equation, i.e., if it represents an eigen-function of this equation, the function $\psi_k(q,t)$ of (2) will be a solution of the second Schrödinger equation (1).

On the other hand, the function $\psi_k(q,t)$ of (2) will then also be a solution of the first Schrödinger equation; this is stated in (3). We see that the first Schrödinger equation leaves its solutions undetermined with respect to a factor which does not depend on q, but may depend on t. It follows that we can consider also the function $\psi_k(q,t)$ as an eigen-function of the energy.

We may therefore consider (2) as a special case of an expansion in eigen-functions of the energy, in which the coefficients σ_m vanish except for σ_k which is $= 1$. In this interpretation, (2) represents case 1. On the other hand, while keeping the idea that all $\sigma_m = 0$ for $m \neq k$, we can also consider $\varphi_k(q)$ as the eigen-function of the energy, and then interpret the exponential factor in (2) as representing the coefficient σ_k. In this interpretation, (2) represents case 2. We comprise both cases in saying that (2) represents a ψ-function which is *given through* an eigen-function of the energy.

Although ψ_k in (2) depends on t, the square $|\psi_k|^2$ is independent of t, since the exponential factor has the absolute value 1 and drops out when we multiply ψ_k by ψ_k^*. Furthermore, in both interpretations $|\sigma_k|^2$ is $= 1$. The probability distributions $d(q)$ and $d(H)$ are therefore time independent. We have a *stationary case*; the ψ-waves represent stationary oscillations which leave observable entities unchanged. Such a case corresponds to one of Bohr's stationary cases of the atom, in which each electron moves in its orbit and the total energy H has a sharp value H_k.

The second Schrödinger equation is so constructed that it follows the principle of superposition. This means that every linear combination of solutions of it is also a solution. We can use this property in order to construct a function $\psi(q,t)$ of a more general form than (2). This is achieved when we write the expansion

$$\psi(q,t) = \sum_k \sigma_k \psi_k(q,t) \tag{5}$$

in which the coefficients σ_k are independent of t, and ψ_k has the meaning (2). The bundle of waves of different frequencies, given in (5), is a solution of (1), because each ψ_k is a solution of (1).

In the form (5) the expansion represents case 1, since it is constructed in time-dependent eigen-functions. With the abbreviation

$$\sigma_k'(t) = \sigma_k e^{-\frac{2\pi i}{h} H_k t} \tag{6}$$

we can write (5) in the form

$$\psi(q,t) = \sum_k \sigma_k'(t) \varphi_k(q) \tag{7}$$

In this form the expansion represents case 2, since we have time-dependent coefficients and time-independent eigen-functions. This double interpretation is possible because the time enters only in the form of an exponential factor which can be included either in the coefficient or in the eigen-function. Since the exponential factor has the absolute amount 1, such an inclusion neither violates the condition (1), § 17, of normalization, nor the condition (2), § 9, of orthogonality.

Both the coefficients σ_k and $\sigma_k'(t)$ are so constructed that their squares $|\sigma_k|^2$ and $|\sigma_k'(t)|^2$ are independent of the time t. We therefore have a probability distribution $d(H)$ of the energy which does not vary with time. It is different with the function $|\psi(q,t)|^2$. This function depends on time, because, when we multiply the sum in (5) or (7) with its complex conjugate, the exponential factor furnishes terms $e^{-\frac{2\pi i}{h}(H_k - H_m)t}$ which do not drop out. We thus have here a probability distribution $d(q)$ which varies with time. The ψ-function (5), or (7), therefore does not represent a stationary case, but a case in which, in the language of Bohr's model, transitions between stationary states occur. The mentioned exponential term indicates, through its exponent, that these transitions are combined with oscillations of the frequency $\dfrac{H_k - H_m}{h}$, in correspondence with Bohr's rule for the emission of radiation. This nonstationary state, however, is represented in (5) as a *superposition of stationary states*, a conception unknown in the older quantum theory.

Reversing the inferences given above, we can prove that if the function $\psi(q,t)$ is expanded in the time-independent eigen-functions $\varphi_k(q)$ of the energy, i.e., if it has the form (7), the coefficients $\sigma_k'(t)$ must have the form (6), and thus will contain the time only within an exponential factor of the amount 1. In other words, the general situation can always be interpreted as a superposition of stationary states, at least if it is a situation for which time-independent eigen-functions of the energy can be defined. The proof is as follows. From (7) we infer the two relations, using in the second line the first Schrödinger equation for the last step:

$$\frac{ih}{2\pi}\frac{\partial \psi}{\partial t} = \sum_k \frac{ih}{2\pi}\frac{d\sigma_k^1}{dt}\varphi_k \tag{8}$$

$$H_{op}\psi = \sum_k \sigma_k' H_{op}\varphi_k = \sum_k \sigma_k' H_k \varphi_k \tag{9}$$

According to (1) the terms of both lines are equal. Comparison of the coefficients of the expansions furnishes

$$\frac{ih}{2\pi}\frac{d\sigma_k'}{dt} = \sigma_k' H_k \tag{10}$$

This differential equation determines σ_k' as being of the form (6).

§ 18. TIME DEPENDENCE 89

This result can be stated in the form: Whatever be the form of the function $\psi(q,t)$, the probability distribution of the energy is time independent.[3] This theorem can be considered as the quantum mechanical generalization of the principle of the conservation of energy. Similar results obtain for other entities which are commutative with the energy, and which, classically speaking, are time-independent integrals of the equations of motion.

If we expand the function $\psi(q,t)$ in terms of the eigen-functions $\chi_m(q)$ of any other entity v, which is not commutative with the energy, we obtain

$$\psi(q,t) = \sum_m \tau_m(t)\chi_m(q) \tag{11}$$

Here the expansion coefficients $\tau_m(t)$ do not have the special form (6), i.e., they do not consist of a constant multiplied by an exponential factor of the amount 1. Their squares $|\tau_m(t)|^2$ will therefore be time dependent. The probability distribution $d(v)$ of an entity v, which is not commutative with the energy, will therefore, in general, vary with time. Only in the stationary case (2), i.e., in a state where ψ is given through an eigen-function of the energy, will the probability distributions $d(v)$ of other entities v be time independent. This follows because the expansion of (2) in the eigen-functions $\chi_m(q)$ of v can be constructed by expanding only $\varphi_k(q)$, and can therefore be written, with constant coefficients τ_m, in the form:

$$\left. \begin{array}{l} \psi_k(q,t) = e^{-\frac{2\pi i}{h} H_k t} \sum_m \tau_m \chi_m(q) \\ \\ = \sum_m \tau_m e^{-\frac{2\pi i}{h} H_k t} \chi_m(q) \end{array} \right\} \tag{12}$$

The exponential factor can here, as before, be included either in the coefficient or in the eigen-function. In both cases the squares of the coefficients are time independent.

The results of this section can be summarized as follows. The state function ψ is always time dependent, i.e., has the form $\psi(q,t)$. Its dependence on time is controlled by the second Schrödinger equation (1). In stationary cases $\psi(q,t)$ consists of a time-independent part $\varphi_k(q)$, which is an eigen-function of the energy, and an exponential factor of the amount 1 containing t. In nonstationary cases of the form considered, the function $\psi(q,t)$, if expanded in terms of the eigen-functions of the energy, has the form (5) or (7), and therefore leads to constant expansion coefficients σ_k, or coefficients $\sigma_k'(t)$ which, according to (6), contain the time only in an exponential factor of the amount 1. The probability function $|\sigma_k|^2$ or $|\sigma_k'(t)|^2$ is therefore time independent. If the nonstationary $\psi(q,t)$ is expanded into the eigen-functions of an entity v which is noncommutative with the energy, the coefficients $\tau_k(t)$ will contain the time

[3] This result is limited to a time-independent energy operator. We shall not enter here into the discussion of more general cases.

in a form different from (6), and the probability distribution $|\tau_k(t)|^2$, therefore, will depend on time.

In the following sections we shall not indicate the variable t in the function ψ, but simply write $\psi(q)$. All relations to be explained then will be understood to hold for every value t.

§ 19. Transformation to Other State Functions

We have so far used the function $\psi(q)$ as state function, i.e., as the function describing the probability state of a physical system. In so doing we have distinguished the coordinates q from other physical entities, since ψ is formulated in terms of the q. The same distinction is expressed in the form of the Schrödinger equation (5), § 14, which determines eigen-functions $\varphi(u,q)$ as relations between any entity u and the coordinates q. We shall now show that this distinction can be eliminated, and that we can use other systems of physical entities in the place of the coordinates q.

Let us consider the continuous case. The considerations to be given then are based on the symmetry of the functions $\psi(q)$ and $\sigma(u)$ as explained in § 9. Applying (16), § 11, we can expand $\sigma(u)$ in terms of the $\varphi(q,u)$, the expansion coefficients being given by the function $\psi(q)$:

$$\sigma(u) = \int \psi(q)\breve{\varphi}(q,u)du \qquad \breve{\varphi}(q,u) = \varphi^*(u,q) \qquad (1)$$

We can interpret this as meaning that $\sigma(u)$ is the state function and that the $\breve{\varphi}(q,u)$ are the eigen-functions of q with respect to u. Using this conception and applying rule II to the expansion coefficient $\psi(q)$, we immediately obtain (14), § 17, i.e., rule III.

Using the geometrical language of the F-space (§ 10), in the passive interpretation, we may say that instead of using the ψ-system as the system of reference we have now introduced the σ-system as the system of reference. The eigen-functions of the entity q in the σ-system are the complex conjugates, with reversed arguments, of the eigen-functions of u in the ψ-system. When we now introduce a third entity v, with a state function $\tau(v)$, we must make use of the triangle of transformations (§ 11). If the eigen-functions of v in the ψ-system are $\chi(v,q)$, we cannot use these functions as the eigen-functions of v in the σ-system; instead we have to introduce eigen-functions $\omega(v,u)$ which are determined by (16e), § 11, as being

$$\omega(v,u) = \int \chi(v,q)\breve{\varphi}(q,u)dq \qquad (2)$$

We see that we must revise our terminology of eigen-functions so far used. We must not speak simply of eigen-functions of a physical entity, but of eigen-functions of a physical entity with respect to a given entity used as argument of the state function.

§ 20. DETERMINATION OF THE Ψ-FUNCTION   91

The transformation of the state function involves some further changes. The second Schrödinger equation, which determines ψ in its dependence on time, must be transformed into a similar equation determining the dependence of $\sigma(u)$ on time. We refer the reader to textbooks on quantum mechanics.[1] With these considerations the absolutism of the coordinates q is to a great extent eliminated and replaced by a relativity of parameters. There is, however, a residue of absolutism in the fact that the transition from the classical function (1), § 14, to the operator of the entity u is defined only for the q and p as arguments.

§ 20. Observational Determination of the Ψ-Function

We saw in § 18 that the second Schrödinger equation determines the ψ-function at any time t, if the ψ-function at the time t_0 is given. In order to make use of this procedure we must therefore know the ψ-function at the time t_0. This determination cannot be made by mere theoretical inquiry; it requires, in addition, experiment and observation.

An analogy with classical mechanics may be used to clarify these considerations. Schrödinger's second equation corresponds to the causal law which determines the change of the values of physical entities in the course of time. In order to apply such laws we must know the initial values of the entities involved; this knowledge is attained by means of experiment and observation. The corresponding knowledge in quantum mechanics is expressed in the ψ-function at the time t_0.

The difficulties in determining the ψ-function originate from the fact that the complex functions of quantum mechanics do not represent immediately observable entities; it is only their squares which symbolize such entities. These squares express probabilities and are therefore accessible to statistical determination. Now the square of a complex function does not determine the function, since many different complex functions may have the same square. The reason is that a complex function represents a pair of two real functions, whereas its square represents only one real function. In order to determine a complex function we must therefore know two suitably chosen real functions of a statistical character.

The results of § 17 suggest an attempt in the following direction. If the function $\psi(q)$ is given, rules III and II determine the probability distributions $d(q)$ and $d(p)$ of position and momentum; the first is given as $|\psi(q)|^2$, the second as $|\sigma(p)|^2$, where $\sigma(p)$ represents the coefficient in the expansion into eigen-functions. We might suspect that if $d(q)$ and $d(p)$ were given, it should be possible to determine both $\psi(q)$ and $\sigma(p)$, since the eigen-functions of the expansion are known; therefore each one of the two functions $\psi(q)$ and $\sigma(p)$ is determined by the other.

[1] Cf., for instance, H. A. Kramers, *Grundlagen der Quantentheorie* (Leipzig, 1938), pp. 145, 153.

It turns out that this supposition is wrong. V. Bargmann, of whom the author asked this question, has shown through the construction of an example that a class of several differing functions $\psi(q)$, and, therefore, also a corresponding class $\sigma(p)$, is compatible with the same distributions $d(q)$ and $d(p)$. It would, of course, be irrelevant if the ambiguity left in $\psi(q)$ were reduced to the occurrence of an arbitrary phase factor, since two functions $\psi(q)$ differing only with respect to a constant phase factor, i.e., a complex factor of the amount 1 which does not depend on q, furnish the same probability distributions for all physical entities. The example constructed by Bargmann, however, shows that the ambiguity in $\psi(q)$ is of a more general kind. Given $d(q)$ and $d(p)$, it is possible to construct two different functions $\psi_1(q)$ and $\psi_2(q)$ compatible with the given distributions, but leading to different distributions $d_1(u)$ and $d_2(u)$ of a third noncommutative entity u.[1]

We therefore must say that the function $\psi(q)$ represents the pair $d(q)$ and $d(p)$ plus some further knowledge. It does not seem possible at the present time to state by what kinds of observation this particular additional knowledge can be acquired.

There exists, however, another way which is free from these ambiguities and which enables us to determine $\psi(q)$ by means of statistical observations. This way has been opened by E. Feenberg[2] who has shown that ψ can be determined if the time dependence of ψ is included in these statistical considerations. As we have pointed out in § 17, the function $\psi(q)$ must be regarded as the special value of a time-dependent function $\psi(q,t)$ for a certain time t. Similarly, the function $|\psi(q)|^2$ represents a special value of a function $|\psi(q,t)|^2$, and we have, therefore, also a time-dependent probability distribution $d(q,t)$. We shall indicate the special functions resulting for the time t_0 by writing t_0 for the argument. Now Feenberg has shown that $\psi(q,t_0)$ is determined to a constant phase factor if the two real functions $d(q,t_0)$ and $\dfrac{\partial}{\partial t} d(q,t_0)$ are given,

---

[1] Bargmann assumes the function $\sigma(p)$ to be of the special form $\sigma(p) = \sigma(-p)$; he then shows that, if $\varphi(p,q)$ is the Fourier function $\dfrac{1}{\sqrt{2\pi}} e^{ipq}$, the two relations hold:

$$\psi(q) = \int \sigma(p)\varphi(p,q)dp$$

$$\psi^*(q) = \int \sigma^*(p)\varphi(p,q)dp$$

This follows, since for the Fourier function $\varphi^*(p,q) = \varphi(-p,q)$, through the relations

$$\psi^*(q) = \int_{-\infty}^{+\infty} \sigma^*(p)\varphi^*(p,q)dp = -\int_{+\infty}^{-\infty} \sigma^*(-p)\varphi^*(-p,q)dp = \int_{-\infty}^{+\infty} \sigma^*(p)\varphi(p,q)dp$$

The function $\psi(q)$ then also satisfies the relation $\psi(q) = \psi(-q)$, but can be arbitrarily chosen apart from this restriction. It follows that the given probability distributions $d(q)$ and $d(p)$ can be combined with the couple ψ, σ as well as with the couple ψ^*, σ^*; these two couples, however, lead to different probability distributions $d(u)$ of other entities u. Starting from this simple example, Bargmann has constructed even more complicated forms of ψ-functions holding for given $d(q)$ and $d(p)$.

[2] Feenberg's proof is presented in E. C. Kemble, *Fundamental Principles of Quantum Mechanics* (New York, 1937), p. 71.

§20. DETERMINATION OF THE Ψ-FUNCTION

the latter expression meaning the time derivative of $d(q,t)$ taken at the time t_0. The proof, which requires some mathematical methods not represented in this book, makes use of the time-dependent Schrödinger equation. This equation establishes, as is seen from (1), § 18, a relation between the shape of the ψ-function in the configuration space and its time derivative, since its left hand side contains space derivatives of ψ and its right hand side represents the time derivative. The knowledge of a time derivative, even if it concerns only the time derivative of the function $|\psi(q,t)|^2$, leads therefore to some determination of the shape of the function $\psi(q,t)$ in the configuration space, which in combination with the value of $|\psi(q,t)|^2$ leads to a complete determination of $\psi(q,t)$, apart from a constant phase factor.

We see that the pair of the two real functions $d(q,t_0)$ and $\frac{\partial}{\partial t} d(q,t_0)$ is superior to the pair $d(q)$ and $d(p)$, or $d(q,t_0)$ and $d(p,t_0)$, so far as only the first pair leads to a determination of the ψ-function. The reason is that the time-dependent Schrödinger equation restricts the choice of ψ-functions compatible with the pair $d(q,t_0)$ and $\frac{\partial}{\partial t} d(q,t_0)$; if we start from the pair $d(q)$ and $d(p)$, there is no such restriction narrowing down the class of possible ψ-functions. Since the momentum p is determined by the velocity, and the latter represents the time derivative of q, we can state our result as follows: The time derivative of the distribution $d(q)$ states more than the distribution of the time derivative of q.

The practical way of determining the function ψ can now be described, at least in principle, as follows. We start, not from one system, but from an assemblage A of systems in equal physical conditions; these conditions must be so specified that they assure the same, though unknown, form of the function $\psi(q,t)$ for each of the systems, or, in other words, that they assure the existence of one ψ-function for the assemblage (cf. § 23). We now select at random a subclass B of the systems and measure at the time t_0 the value of q in each of the systems of the subclass B. The results will vary with the individual systems, and we thus obtain a probability distribution $d(q,t_0)$. We then assume that the so-obtained distribution holds equally for the remainder of the class A, i.e., we assume that we would have obtained the same distribution if we had measured q in all systems of A. This assumption represents the usual inductive inference without which, of course, no physical statements could be constructed.

We now make an important logical step. The systems of the subclass B, since a measurement has been performed on them, have been disturbed, and they are therefore excluded from the assemblage. These systems now have new ψ-functions which do not interest us; the ψ-function whose square is represented by the obtained distribution $d(q,t_0)$ is the ψ-function of the remainder assemblage A'.

94   PART II. MATHEMATICAL OUTLINES

At the time t_1 shortly after t_0 we select another subclass B' from A', and now make measurements of position on all the systems of B'. The resulting distribution $d(q,t_1)$ is then considered as determining the square of the function $\psi(q,t_1)$ of the remainder assemblage A'', resulting from A' by the exclusion of B'. The difference $d(q,t_1) - d(q,t_0)$, divided by $t_1 - t_0$, then represents a numerical approximation for the derivative $\frac{\partial}{\partial t} d(q,t_0)$, or $\frac{\partial}{\partial t} |\psi(q,t_0)|^2$.

We now must determine a function $\psi(q,t_0)$ which satisfies the two given numerical data $|\psi(q,t_0)|^2$ and $\frac{\partial}{\partial t}|\psi(q,t_0)|^2$, and which at the same time satisfies the time-dependent Schrödinger equation. This determination, since it is to lead only to approximative results, can be carried out by numerical methods. We thus finally arrive at a function $\psi(q,t_0)$ which we consider, not only as the ψ-function of the assemblage A'', but of any assemblage of the kind A, i.e., of any assemblage produced under the physical conditions by which A was defined.

The value of this consideration, although it cannot always be carried through for practical reasons, consists in the fact that it shows the way to an observational determination of the ψ-function. *The state function ψ can be ascertained in terms of observational data;* this is what we want to show here. Once this is made clear it appears advisable to look also for other ways of determining the ψ-function. A frequently used method is to guess the form of the ψ-function from the general physical conditions of the arrangement, and to check its numerical constants from observational results derivable from the assumed function. Such methods are permissible after it has been made clear to what extent the observational data used determine the ψ-function.

We now see the reason that complex functions are used in quantum mechanics. The complex function ψ can be considered as a mathematical abbreviation which stands for pairs of statistical sets. Thus, we can consider $\psi(q,t_0)$ as the representative of the pair $d(q,t_0)$ and $\frac{\partial}{\partial t} d(q,t_0)$; or, since the time derivative is determined by the distribution at an immediately following time t_1, as the representative of the pair of two consecutive distributions $d(q,t_0)$ and $d(q,t_1)$; or, as the representative of the pair $d(q,t_0)$ and $d(p,t_0)$, with the qualification that ψ contains some more indications than the latter pair. The value of the mathematical algorism of quantum mechanics consists in the fact that with the knowledge of the complex function $\psi(q)$ we know, not only the two original statistical sets from which $\psi(q)$ was derived, but also the statistical distributions of all other entities of the system at the same time; and, furthermore, the statistical distributions of all these entities at any later time. The procedure by which this is accomplished is given in the rules I and II, § 17, and in the time-dependent Schrödinger equation.

A remark of a logical nature must be added. Since the determination of a ψ-function is achieved by statistical methods, a ψ-function must be ascribed, not to an individual system, but to an assemblage. Now the same individual system may belong to different assemblages; therefore, expressions like "the ψ-function of this system" are, strictly speaking, meaningless. In spite of this difficulty it is permissible to use such expressions when it is clear, from the context of the consideration, to which assemblage the system is referred. A similar difficulty is known for statements about the degree of probability of a single case; since probability is a property definable only for classes, such statements, too, are meaningful only when the case under consideration is conceived as incorporated into a certain reference class.[3] When this class is not specifically named it must be required that it be evident from the context. Similarly, the term "ψ-function of a certain physical system" will be considered by us as meaningful when it is clear which *reference class* is understood.

§ 21. Mathematical Theory of Measurement

We now shall turn to an analysis of *measurement*. We have made use of measurements of the values of entities in our statistical consideration; and we must discuss what is to be meant by a measurement.

When in classical physics we distinguish between a measured and an unmeasured entity, the difference of the two cases can be analyzed as follows. An unmeasured entity is not known; but that does not exclude that it can be predicted with a certain probability. So long as this probability is smaller than 1, we must say that the entity is unknown. In order to know the entity, we have to perform further physical operations by which the entity is measured. After these operations have been made, the uncertainty of the prediction has been overcome; this means we now can predict with certainty the result we would obtain if we were to repeat the measurement.

We can use this idea for a definition of measurement:[1] *A measurement is a physical operation which furnishes a determinate numerical result and which in an immediate repetition furnishes the same result.*

The addition of the term "immediate" is necessary, because, if we wait too long, the entity may change its value. Only for time-independent entities can this qualification be omitted.

It is obvious that the use of the term "measurement" in classical physics is covered by our definition, in the sense that every classical measurement has the property required in our definition. Usually we combine with the word "measurement" in classical physics some further requirements, such as the requirement that the entity be not produced by the measurement but be in existence before it, etc. Let us postpone the discussion of such questions and

---

[3] Cf. the author's *Experience and Prediction* (Chicago, 1938), §§ 33–34.
[1] This definition of measurement has been very clearly formulated by E. Schrödinger, *Naturwissenschaften* 23 (1935), p. 824.

keep to the given definition, if for no other reason than that, for our present analysis, it frees us from all such questions.

It is an advantage of the given definition that it can be applied also in quantum mechanics. In order to show this we must first give a translation of our definition into the mathematics of quantum mechanics.

Since the probability state of a physical system is expressed through the ψ-function, our definition requires the introduction of a ψ-function which permits us to make strict predictions, i.e., predictions relative to which the probability distribution $d(u)$ of an entity u degenerates into a concentrated distribution. We use the term *concentrated distribution* in the discrete case for a distribution such that $d(u_i) = 1$ for the subscript i, but $d(u_k) = 0$ for all other subscripts $k \neq i$. For the continuous case the concentrated distribution can only be defined in terms of approximations; the term then means that $d(u)$ can be considered as a Dirac function $\delta(u, u_1)$, where u_1 is the existing value of the entity u.

Now it is easy to see that, in the discrete case, $d(u)$ is a concentrated distribution when the function $\psi(q)$ is identical with an eigen-function $\varphi_i(q)$ of the entity u. In the term "identical" we here include the possibility that $\psi(q)$ and $\varphi_i(q)$ differ by a constant phase factor of the amount 1. The theorem stated follows, because the expansion of $\psi(q)$ in terms of the $\varphi_k(q)$, namely,

$$\psi(q) = \varphi_i(q) = \sum_k \sigma_k \, \varphi_k(q) \tag{1}$$

furnishes in this case the relations

$$|\sigma_i| = 1 \qquad \sigma_k = 0 \quad \text{for} \quad k \neq i \tag{2}$$

According to rule II, § 17, this means

$$P(s, u_1) = 1 \qquad P(s, u_k) = 0 \quad \text{for} \quad k \neq i \tag{3}$$

For the continuous case we have here some mathematical complications. Since continuous basic functions are not reflexive (cf. § 9), the function $\psi(q)$ cannot be identical with one of the eigen-functions of u. Using, however, methods of approximation, we can carry through considerations analogous to those of the discrete case. Instead of a sharp value u_i we consider here a small interval $u_1 \pm \epsilon$, which we abbreviate as u_ϵ. The condition that $d(u)$ be a concentrated distribution then is given by the relations expressing an approximate equality

$$P(s, u_\epsilon) \sim 1 \qquad P(s, u) \sim 0 \quad \text{for } u \text{ outside } u_\epsilon \tag{4}$$

This condition is satisfied if

$$\int_{u_1-\epsilon}^{u_1+\epsilon} |\sigma(u)|^2 \, du \sim 1 \qquad \sigma(u) \sim 0 \quad \text{for } u \text{ outside } u_\epsilon \tag{5}$$

§21. THEORY OF MEASUREMENT

The expansion of ψ then is practically given by

$$\psi(q) \sim \int_{u_1-\epsilon}^{u_1+\epsilon} \sigma(u)\varphi(u,q)du \tag{6}$$

This means that $\psi(q)$ is composed practically of a bundle of contiguous eigen-functions $\varphi(u,q)$ corresponding to values u within the interval u_ϵ. We shall say that in this case $\psi(q)$ is a *practical eigen-function* of the entity u.[2] These results enable us to give the following definition.

Quantum mechanical definition of measurement: A measurement of an entity u is a physical operation relative to which the ψ-function of the physical system is represented by one of the eigen-functions of u, or by one of the practical eigen-functions of u.

This definition must be considered as one of the fundamental principles of quantum mechanics; it expresses the part allotted to measurement in a world of probability connections. That there are physical arrangements which are measurements in the sense defined, is a matter of fact; the description and construction of such arrangements constitute that part of physics which we call the technique of measurement. We shall call a measurement defined in terms of eigen-functions, as presented in the above definition, an *ideal measurement*. It follows from the definition of practical eigen-functions that in the continuous case we have only more or less ideal measurements, i.e., measurements that can be made only to a certain degree of approximation. In the discrete case an ideal measurement can be made in principle; but in practice this will not be possible, and even in this case we shall therefore have a ψ-function which is only a practical eigen-function of the entity measured, i.e., which will be composed of a bundle of contiguous eigen-functions. In both these cases we shall call the system *definite in u*.[3]

We now can turn to the discussion of the role played by measurements with-

---

[2] It should be realized that the values of the coefficients $\sigma(u)$ inside the interval u_ϵ are left completely undetermined by the requirement (5), and that therefore the shape of the function $\psi(q)$ is not determined. What we call a practical eigen-function is therefore not to be imagined as a function similar to an eigen-function. It is true that (6) can be written approximately as

$$\psi(q) \sim \varphi(u_1,q) \int_{u_1-\epsilon}^{u_1+\epsilon} \sigma(u)du$$

But the integral in this expression will converge toward zero with smaller ϵ, whereas $\varphi(u,q)$ is finite; this convergence is necessary because $\psi(q)$ is normalized according to (1), § 17, whereas the corresponding integral for $\varphi(u,q)$ is infinite. $\psi(q)$ then will also converge toward zero, whereas the condition of normalization remains satisfied. This means that the standard deviation Δq occurring in Heisenberg's inequality (2), § 3, goes toward infinite values. Since the order of magnitude of $\psi(q)$ will then be the same as that of the unexactness of the above relation, it is not possible to say that $\psi(q)$ is proportional to $\varphi(u,q)$. It follows that the process of going to the limit $\epsilon = 0$ does not determine a definite form of the ψ-function. The continuous case differs here essentially from the discrete case.

[3] It follows from the considerations given in the above footnote that if a system is definite in u its ψ-function is not yet determined, and that therefore the probability distributions of other entities will not be determined. Only if we have the discrete case and know that $\psi(q)$ is strictly equal to an eigen-function $\varphi_i(u)$ are the probability distributions of all other entities also determined. In general, we need two probability distributions, as explained in § 20, in order to determine $\psi(q)$.

in physical operations. We shall base our discussion on the ideal definition of measurement given above. We shall see that this definition leads immediately to the relation of indeterminacy and to a formulation of the disturbance emanating from the measurement. For the ideal measurement this disturbance will represent a minimum; actual measurements will always produce greater disturbances. This fact justifies a restriction of the discussion to an analysis of ideal measurements; we then determine the minimum of disturbance which any measurement can produce.

The relation of indeterminacy concerns the question of different measurements made at the same time, i.e., by the same physical arrangement. If such an arrangement is to determine the exact values of different physical entities, these entities must have the same eigen-functions; otherwise the ψ-function of the arrangement can correspond only to the eigen-function of one of the entities. We showed above that entities with commutative operators have the same eigen-functions; therefore, such entities can be measured simultaneously. Entities with noncommutative operators, however, have different eigen-functions; it is therefore impossible to measure such entities by the same physical arrangement. We saw in (6), § 15, that, in particular, momentum and position are noncommutative entities; it follows that there exists no physical arrangement which measures simultaneously position and momentum.

If these entities, however, cannot be measured simultaneously, is it not possible to measure one after the other? We must turn to the answer of this question. For the sake of simplicity we shall deal here only with the heterogeneous case; the continuous case can be reduced to the same results by approximative methods such as explained before.

If a measurement of u has been made with the result u_i, the ψ-function of the system is identical with $\varphi_i(q)$. In order to determine the probability of measuring, in such a system, a value of an entity v, we must expand this ψ-function into the eigen-functions of v. Since v is assumed to be noncommutative with u, these eigen-functions $\chi_k(q)$ are different from the $\varphi_i(q)$. Now the expansion of $\varphi_i(q)$ in terms of the $\chi_k(q)$ is determined by the triangle of transformations (§ 11); and (17f), § 11, furnishes

$$\varphi_i(q) = \sum_k \breve{\omega}_{ik} \chi_k(q) \qquad (7)$$

We now can apply rule II, § 17. Since the situation s of the system is a situation produced by a measurement of u with the result u_i, we can replace s by u_i in the probability expression, and write, by analogy with (3), § 17:

$$P(u_i, v_k) = |\breve{\omega}_{ik}|^2 \qquad (8)$$

This is the probability of finding the value v_k after the value u_i has been measured. We see that we have here, not a concentrated distribution, but a probability spectrum extending over all possible values of v. Predictions concerning

§21. THEORY OF MEASUREMENT

the measurements of v after a measurement of u has been made, can therefore be made only with determinate probabilities.

Now imagine that after the measurement of u with the result u_i, a measurement of v is actually made, and furnishes the result v_k. We then have a new physical situation s with a new ψ-function, namely, a ψ-function which is identical with $\chi_k(q)$. We know that, if we were to repeat the measurement of v, we would once more obtain v_k; this is shown by considerations similar to those used before. But if we want to make, instead, a new measurement of u, we can no longer predict its result with certainty. Rather, we have a probability spectrum for u, given by the expansion of the ψ-function $\chi_k(q)$ in terms of the eigen-functions $\varphi_i(q)$. This expansion is determined by (17d), §11:

$$\chi_k(q) = \sum_i \omega_{ki}\varphi_i(q) \qquad (9)$$

We therefore have

$$P(v_k, u_i) = |\omega_{ki}|^2 \qquad (10)$$

With (2), §11, we have

$$|\breve{\omega}_{ik}|^2 = |\omega_{ki}^*|^2 = |\omega_{ki}|^2 \qquad (11)$$

and therefore

$$P(u_i, v_k) = P(v_k, u_i) \qquad (12)$$

Relation (10) expresses the fact that after the measurement of v we have only a probability spectrum for u. The symmetry of this spectrum with that existing for v after the measurement of u, expressed in (12), does not eliminate the fundamental fact that we cannot get rid of probabilities here. Before the measurement of v we knew with certainty that a second measurement of u would have furnished the value u_i, and we knew with the probability (8) that a measurement of v would produce the value v_k. After the measurement of v with the result v_k we know with certainty that a second measurement of v would furnish v_k, and we know with the probability (10) that a measurement of u would produce the value u_i.

It is this fact which must be interpreted as a disturbance of the object by the measurement. There are measurements in quantum mechanics; this means, there are physical operations with numerical results which, if repeated, lead to the same result. But if two measurements of an entity u are separated by a measurement of a noncommutative entity v, the second measurement of u is not bound to reproduce the previous value u_i, and its result can be foretold only with the probability (10). The measurement of v must therefore have effected a physical change in the conditions of the system.

Mathematically speaking, the influence of the measurement finds its expression in the commutation rule, which (6), §15, formulates for the fundamental operators p and q, and which holds similarly for a series of other parameters. It is the noncommutativity of the operators which makes it impossible to create a physical situation which represents a simultaneous measurement of such parameters. The commutation rule expresses, therefore, in a general form, the same idea which we formulated above as the principle of inverse correlation. It thus can be considered as a second formulation of Heisenberg's principle of indeterminacy, (2), § 3, the latter inequality being derivable from the commutation rule, as is shown in the textbooks of quantum mechanics. If h were $= 0$, all parameters would be commutative, and we would have no disturbance by the measurement and no indeterminacy.

Let us add a remark about the fundamental parameters p and q. The relation *canonically conjugated*, or *complementary*, divides these parameters into two classes such that each class contains only commutative parameters; but this division can be given in various ways. Thus, we can put into one class all the q_i, into the other class all the p_i. We can, however, also put into the first class some of the position parameters and some of the momentum parameters, if the latter are not complementary to the first. Thus, for a mass particle with three position coordinates q_1, q_2, q_3, and the momentum parameters p_1, p_2, p_3, we can put into the first class the entities q_1, p_2, q_3, and into the second class the entities p_1, q_2, p_3. For each such class a compound measurement can be devised such that the parameters of this class are determined; then the parameters of the complementary class remain undetermined.

In a similar way, as holds for momentum and position, energy and time are coordinated to each other as complementary parameters. Formally speaking, there is a difference so far as time is not considered in quantum mechanics as a physical entity of the same kind as other physical entities; thus, there is no time operator to be used in quantum mechanical equations. In consequence of this fact a commutation rule of the form (6), § 15, cannot be established for energy and time. It can even be shown that, if an attempt were to be made to introduce a time operator, a commutation rule of the form (6), § 15, for energy and time would lead to contradictions.[4] In spite of that, the uncertainty relation (3), § 3, holds; but it possesses no analogue in a commutation rule.

A measurement of the energy is, in general, incompatible with a measurement of position or momentum; thus, if the energy operator has the form (3), § 14, it is noncommutative with the position operator as well as with the momentum operator. A measurement which determines the maximal number of parameters that can be ascertained in one measurement is called a *maximal measurement*. Of course there are also measurements which determine a smaller number of entities; for instance, we can measure q_1 alone.

§ 22. The Rules of Probability and the Disturbance by the Measurement

The influence of the measurement can also be stated by means of another consideration. Let us assume that first a measurement m_u of an entity u is made,

---

[4] Cf. W. Pauli, "Die allgemeinen Prinzipien der Wellenmechanik," *Handbuch der Physik*, Vol. XXIV, 1 (ed. by Geiger-Scheel, 2d ed., Berlin, 1933), p. 140.

§22. DISTURBANCE BY THE MEASUREMENT

then a measurement m_v of an entity v, and then a measurement m_w of an entity w. We then have probabilities

$$P(u_i, v_k) \quad \text{and} \quad P(v_k, w_m) \tag{1}$$

determining the results of the measurements. The probability of obtaining a value w_m then depends only on the value v_k obtained, but is independent of the value u_i which preceded the measurement of v, i.e., we have

$$P(u_i.v_k, w_m) = P(v_k, w_m) \tag{2}$$

if the measurements are made in the order which we stated. The period sign within the parentheses denotes the logical "and". (2) follows, because, with the measurement of v, the ψ-function has assumed the form $\chi_k(q)$ which contains no reference to the preceding measurement of u. We therefore have, according to the rules of probability,[1] starting with the general theorem of multiplication:

$$P(u_i, v_k.w_m) = P(u_i, v_k) \cdot P(u_i.v_k, w_m)$$
$$= P(u_i, v_k) \cdot P(v_k, w_m) \tag{3}$$

where $P(u_i, v_k.w_m)$ means the probability that after finding u_i we shall find v_k *and* then w_m. Applying a theorem which in the calculus of probability is known as the *rule of elimination*, we can calculate the probability that u_i is followed by any value of v, and then by w_m, as

$$P(u_i, [v_1 \lor v_2 \lor \ldots].w_m) = \sum_k P(u_i, v_k) \cdot P(v_k, w_m) \tag{4}$$

The sign $\lor$ denotes the logical "or". The left hand side is the same as

$$P(u_i, w_m) \tag{5}$$

since one of the values $v_1, v_2 \ldots$ must occur. The first term on the right hand side of (4) is given by (8), §21. In order to interpret the second term we must realize that we have here a quadrangle of transformations (figure. 8, §11) in which ψ represents the entity q, σ the entity u, τ the entity v, and ρ the entity w. Using the triangle $\psi\tau\rho$ (or qvw) we therefore have, as stated in (27), §11:

$$\chi_k(q) = \sum_m \eta_{km} \xi_m(q) \tag{6}$$

and therefore

$$P(v_k, w_m) = |\eta_{km}|^2 \tag{7}$$

---

[1] The notation which we use here for probability expressions and the rules of operations with such terms are presented in the author's *Wahrscheinlichkeitslehre* (Leiden, 1935). The symbol W used there is here replaced by P, in adaptation to the English language. The same replacement has been carried through in the author's paper, "Les fondements logiques du calcul des probabilités," *Ann. de l'Inst. Henri Poincaré*, tome VII, fasc. V, pp. 267–348. This paper contains a summary of the author's construction of the calculus of probability.

Consequently (4) can be written

$$P(u_i, w_m) = \sum_k |\breve{\omega}_{ik}|^2 \cdot |\eta_{km}|^2 \qquad (8)$$

On the other hand, we can also omit the measurement of v and pass directly from the measurement of u to that of w. In the mathematical algorism this means that we expand the function $\psi(q) = \varphi_i(q)$ in the eigen-functions $\breve{\zeta}_m(q)$ of the entity w. The triangle $\sigma\rho\psi$ (figure 8, § 11) furnishes for this expansion the formula (26), § 11, which we write here in the variable q:

$$\varphi_i(q) = \sum_m \vartheta_{im} \breve{\zeta}_m(q) \qquad (9)$$

We therefore have:

$$P(u_i, w_m) = |\vartheta_{im}|^2 \qquad (10)$$

According to (19), § 11, the triangle $\sigma\rho\tau$ determines ϑ_{im} as

$$\vartheta_{im} = \sum_k \breve{\omega}_{ik} \cdot \eta_{km} \qquad (11)$$

and therefore we have

$$P(u_i, w_m) = \left| \sum_k \breve{\omega}_{ik} \cdot \eta_{km} \right|^2 \qquad (12)$$

The relation (12) contradicts the relation (8), since, in general, the square of the sum is different from the sum of the squares. Stating it more precisely, we may say: For most entities w the relation (12) contradicts (8).

Now the rules of probability are tautological if the frequency interpretation of probability is assumed;[2] therefore we cannot exclude the use of (4), the rule of elimination. The rules of logic cannot be affected by physical experiences. If we express this idea in a less pretentious form, it means: If a contradiction arises in physical relations, we shall never consider it as due to formal logic, but as originating from wrong physical interpretations. In our case the mistake is given in the expressions used on the left hand sides of (8) and (12); these two expressions do not mean the same, and should therefore be distinguished by a suitable notation.

For this purpose we must indicate the fact that a measurement has been made. Let us denote a measurement of u by m_u. Instead of (8), § 21, we then shall write:

$$P(m_u.u_i.m_v,v_k) = |\breve{\omega}_{ik}|^2 \qquad (13)$$

Here the order of the terms connected by the period sign expresses the time order.[3] The relation (12), § 21, will be written in our new notation:

$$P(m_u.u_i.m_v,v_k) = P(m_v.v_k.m_u,u_i) \qquad (14)$$

---

[2] Cf. the author's *Wahrscheinlichkeitslehre* (Leiden, 1935), § 18.
[3] We are therefore using in this notation an asymmetrical "and". This can be avoided by the use of superscripts expressing the time order.

§22. DISTURBANCE BY THE MEASUREMENT

For (7) we now have:
$$P(m_v.v_k.m_w,w_m) = |\eta_{km}|^2 \tag{15}$$

For (8) we shall now write:
$$P(m_u.u_i.m_v.m_w,w_m) = \sum_k |\breve{\omega}_{ik}|^2 \cdot |\eta_{km}|^2 \tag{16}$$

and (12) assumes the form
$$P(m_u.u_i.m_w,w_m) = \left| \sum_k \breve{\omega}_{ik} \cdot \eta_{km} \right|^2 \tag{17}$$

Instead of a contradiction we then simply derive the inequality
$$P(m_u.u_i.m_v.m_w,w_m) \neq P(m_u.u_i.m_w,w_m) \tag{18}$$

This inequality states clearly the influence of the measurement: The probability of obtaining a value w_m does not depend only on *results* of preceding measurements of other entities, but also on the *fact* of such measurements. The term m_v entering into the first place of the probability expression changes the probability, even if no value v_k occurs in this expression.

The amount of this influence of the measurement m_v can be exactly calculated so far as probabilities are concerned; this is expressed in (16). That, in spite of the various forms in which an entity can be measured, such a general calculation of the influence of the measurement can be given is due to the fact that the theory is concerned only with ideal measurements (cf. p. 97). For actual measurements (16) will furnish only approximative results.

The influence of the measurement is even more apparent if we choose as third entity w the first entity u, i.e., if we repeat the first measurement. Without an intervening measurement of v we then have (cf. (3), §21):[4]

$$P(m_u.u_i.m_u,u_m) = \delta_{im} \tag{19}$$

This means that we have a concentrated system; the probability of getting the same value as observed before is = 1, and the probability for any other value is = 0. However, when we intercalate a measurement of v we have, with (4), and using (8), §21, and (10), §21,

$$P(m_u.u_i.m_v.m_u,u_m) = \sum_k |\breve{\omega}_{ik}|^2 \cdot |\omega_{km}|^2 \tag{20}$$

This is a general distribution, as though there had been no preceding measurement of u. The intervening measurement of v has destroyed the concentrated distribution existing for u before, and we arrive at the inequality corresponding to (18):
$$P(m_u.u_i.m_v.m_u,u_m) \neq P(m_u.u_i.m_u,u_m) \tag{21}$$

The physical disturbance by a measurement is of such a kind that every

---

[4] (19) follows also from (12) by the substitution of ω_{km} for η_{km} and application of (2), §11, and the orthogonality condition for ω in the form (7), §9.

measurement creates a new situation in which the influence of the preceding situation is no longer recognizable. This fact is expressed by the two relations

$$P(s.m_v.v_k.m_w,w_m) = P(m_v.v_k.m_w,w_m) \qquad (22)$$

$$P(m_u.u_i.m_v.v_k.m_w,w_m) = P(m_v.v_k.m_w,w_m) \qquad (23)$$

which state that the term s, and also terms like $m_u.u_i$, can be dropped in the first place of a probability expression if they are followed by terms like $m_v.v_k$. These relations take the place of the incorrectly written relation (2).

The inequalities (18) and (21) constitute a very clear formulation of the disturbance of the object by the measurement. We can use this formulation for a proof of our statement that the principle of indeterminacy (cf. p. 17) is not a logical consequence of the disturbance by the observation. For this purpose we must show that a physical world is possible in which the measurement disturbs the object, but in which a strict prediction of the result of the measurement can be made. Such a world can be constructed as follows. Imagine that the relation holds:

$$P(m_u.u_i.m_v,v_k) = \delta_{i+1,k} \qquad (24)$$

This means that the measurement always changes the system in such a way that the eigen-value with the next higher subscript is produced. Let us assume that this relation holds also if we put m_u for m_v, and u_k for v_k, i.e., if the measurement of the entity u is repeated. We then obtain with the rule of elimination:

$$P(m_u.u_i.m_v.m_w,w_m) = \sum_k P(m_u.u_i.m_v,v_k) \cdot P(m_v.v_k.m_w,w_m)$$

$$= \sum_k \delta_{i+1,k} \cdot \delta_{k+1,m} = \delta_{i+2,m} \qquad (25)$$

whereas we have with (24)

$$P(m_u.u_i.m_w,w_m) = \delta_{i+1,m} \qquad (26)$$

This result shows that, as before, the inequality (18) holds, and that we therefore must speak here of a disturbance of the object by the measurement. (24), however, shows that the result of each measurement can be strictly predicted.

(24) is, of course, not valid in quantum mechanics. We use this formula only for the formal purpose of proving that the principle of indeterminacy is not a logical consequence of the disturbance of the object. It is rather the reverse relation which holds. If there were no disturbance by the measurement, we could make strict predictions; reversing this implication by negating implicans and implicate,[5] we come to the result that the principle of uncertainty implies a disturbance of the object by the measurement.

---

[5] This is the logical principle of contraposition formulated by us in (9), § 31.

§ 23. NATURE OF PROBABILITIES 105

§ 23. The Nature of Probabilities and of Statistical Assemblages in Quantum Mechanics

Although the difference between the right hand sides of (16), § 22, and (17), § 22, and of (19), § 22, and (20), § 22, has always been correctly interpreted as expressing the disturbance through measurements, this result has sometimes been obscured by an inappropriate terminology. The phrase "interference of probabilities" has been used to express the fact that complex numbers are used, in the derivation of probabilities in (17), § 22, in a way resembling the mathematical treatment of the interference of waves. This phrase, however, can be understood as meaning that quantum mechanics uses a calculus of probability different from the classical one. We see that this is not the case. It is *probability amplitudes*, not probabilities, that interfere. We know from the considerations of § 20 that such amplitudes represent, not probability functions, but pairs of probability functions; we therefore should not be surprised if laws combining such pairs of probability functions are structurally different from laws governing individual probability functions. Furthermore, since each such pair characterizes a specific physical situation, the laws controlling the relations of such pairs determine the probability values for various kinds of physical situations; they therefore include statements about the influence of physical situations on the probability values pertaining to them. Once a statement about probabilities is derived in quantum mechanics it has the same meaning as ordinary probability statements, and nowhere is there any deviation from the classical rules of probability. A correct notation makes this clear; in following the rules of the calculus of probability it will point out the physical content of quantum mechanical statements. What distinguishes quantum mechanics from classical physics is the inequality (18), § 22, which states that the probability of obtaining a certain result of a measurement depends on the fact of other measurements. This means a difference in physics, but not in the calculus of probability.

The origin of such misinterpretations is perhaps to be found in a misinterpretation of the calculus of probability, which, throughout the history of this calculus, has greatly contributed to the growth of erroneous conceptions about the nature of probability. The calculus of probability is concerned only with the derivation of probabilities from other probabilities; it can therefore be applied for practical purposes only if some probabilities are given. The determination of these initial probabilities is not a mathematical, but a physical problem. In most cases this determination is given simply by the use of the *statistical inference*—also called *induction by enumeration*, or *a posteriori determination* of probabilities; we then count a frequency within an observed sequence of events and assume this frequency to persist for further prolongations of the sequence. We speak here of *statistically inferred probabilities*. There

106   PART II. MATHEMATICAL OUTLINES

are, however, other cases in which we introduce probabilities by means of theoretical assumptions which are not immediately based on statistical considerations. Such assumptions may concern individual cases, or they may be established within the frame of physical theories; i.e., they may have the form of physical laws stating that under certain physical conditions a probability has this or that value. Examples of such *theoretically introduced probabilities* are given by the statement that the probability of a given face of a die is $\frac{1}{6}$, or that all arrangements of molecules in a Boltzmann-Maxwell gas are equally probable. Such statements sometimes have been considered as derivable *a priori* by means of a "principle of no reason to the contrary", which was assumed to constitute a part of the calculus of probability. This is, of course, an untenable conception. The justification of theoretically introduced probabilities can be given only in the way in which any other physical assumption is verified: by testing its observable consequences. The quantum mechanical probabilities, defined in rules II and III, § 17, as resulting from probability amplitudes, are theoretically introduced probabilities in the sense defined. The probability amplitudes, however, do not belong to the calculus of probability, but to physics. The justification of this derivation of probabilities is given, as in all other cases of derivations of observable numerical values, by the success of the assumption. The hypothetical manner of their introduction, however, does not make them probabilities of another kind. Expressions like "potential probabilities", sometimes used to distinguish these probabilities from other probabilities, must appear inadequate. All probabilities are potential so far as they determine the results of future observations. The meaning of probabilities, whether they are introduced by the statistical inference or by way of physical assumptions, is always given in a statement about the limit of a frequency in a sequence.

Keeping to the results of these general considerations, we now can understand a physical distinction which must be made with respect to statistical assemblages occurring in quantum mechanics, known as the distinction between the *pure case* and the *mixture*. The nature of this distinction may be made clear by the following consideration.

Imagine we have a great number of systems in which we make a measurement of u; some systems will furnish a value u_i, others a value u_k, etc. Let us assume that the eigen-values $u_i, u_k, \ldots$ are not degenerate, i.e., that each of these eigen-values corresponds to a specific eigen-function. We then collect in one class C all those systems in which we obtained u_i; this assemblage is said to constitute a *pure case*. It is so called because if we were to repeat our measurement of u in this class we should find u_i in each of the systems. Now imagine we make a measurement of a noncommutative entity v on all systems of C; we then shall obtain in one system the value v_k, in another the value v_m, etc., the probabilities of these values being determined by (13), § 22. After this

§23. NATURE OF PROBABILITIES

measurement the assemblage C no longer constitutes a pure case, but a *mixture*, since it contains systems with different values of v. If we now repeat the measurement of u, we shall find different values of u in the various systems; the percentage of a value u_m in C is then determined by the probability (20), § 22. The pure case, once destroyed, therefore cannot be reëstablished through the performance of measurements alone; in addition, we must make a new selection out of C in order to construct a new pure case.

These considerations will serve to clarify a terminology frequently used with respect to individual systems. When we say that a measurement *on an individual system* creates a pure case, or reduces the ψ-function of the system to an eigen-function of the measured entity, we mean that such will be the ψ-function of an assemblage constructed by putting into one class the system considered *and* a number of other systems with the same result of the measurement. The latter class constitutes the *reference class* (cf. the remarks at the end of § 20) which is understood in the expression "ψ-function of an individual system after a measurement". This phrase, therefore, means the same as "ψ-function of a class of systems after a measurement of u with the result u_i".

Whereas the assemblage representing a pure case consists of systems each having the same ψ-function, namely, a ψ-function given by the eigen-function $\varphi_i(q)$, the subscript i being the same for all systems, the mixture includes systems of various forms of ψ-functions, namely, functions $\varphi_k(q)$ with various subscripts k. Each function $\varphi_k(q)$ will be realized in a certain percentage c_k of cases. Now both assemblages are only statistical assemblages, since even in the pure case the results of measurements of an entity v having an eigen-function different from the ψ-function can only be statistically predicted. The question arises: Is it possible to characterize the mixture, by analogy with the pure case, in terms of one function $\psi(q)$? This function, of course, would have to be different from each of the eigen-functions $\varphi_k(q)$.

The answer is negative; i.e., it is not possible to characterize a mixture in terms of one ψ-function. We prove this by assuming that there is one such ψ-function and showing that this assumption leads to contradictions of the kind explained in § 22. Let us denote the mixture by S. If there is a ψ-function of the mixture S, it can be expanded in the eigen-functions $\varphi_k(q)$:

$$\psi(q) = \sum_k \sigma_k \varphi_k(q) \tag{1}$$

According to rule II, § 17, the value $|\sigma_k|^2$ represents the probability of measuring the value u_k of the entity u; since this result will be found in all those cases in which the individual system is represented by the eigen-function $\varphi_k(q)$, we have

$$P(S.m_u, u_k) = |\sigma_k|^2 = c_k \tag{2}$$

Now let us consider the question of measurements, not of the entities u or v, but of another entity w with the eigen-functions $\breve{\zeta}_m(q)$. Writing ψ in these eigen-functions we have with (24), §11:

$$\psi(q) = \sum_m \rho_m \breve{\zeta}_m(q) \qquad (3)$$

and therefore with the use of (25), § 11:

$$P(S.m_w,w_m) = |\rho_m|^2 = \left|\sum_k \sigma_k \vartheta_{km}\right|^2 \qquad (4)$$

On the other hand, this probability can be determined as follows. Writing φ_k in the eigen-functions $\breve{\zeta}_m$, we have with (26), § 11:

$$\varphi_k(q) = \sum_m \vartheta_{km} \breve{\zeta}_m(q) \qquad (5)$$

Since the probability of having the eigen-function $\varphi_k(q)$ in the mixture S is given by c_k, we have

$$P(S.m_w,w_m) = \sum_k c_k \cdot |\vartheta_{km}|^2 = \sum_k |\sigma_k|^2 \cdot |\vartheta_{km}|^2 \qquad (6)$$

We are led here to the same kind of contradiction as presented by (8), § 22, and (12), § 22. Although there are exceptional matrices ϑ_{km} which satisfy both (4) and (6), most unitary matrices ϑ_{km} will not do so; in other words: It is always possible to construct a unitary matrix ϑ_{km}, or, what is the same, to define an entity w, for which (4) and (6) are not compatible.

It follows that it is not permissible to consider the assemblage S as characterized by one ψ-function. The assemblage S can be characterized only by a mixture of ψ-functions. Using the notation introduced in the probability expressions (16), § 22, and (17), § 22, we can characterize the assemblage S by saying that measurements of w in S have the form $m_u.u_i.m_v.m_w$, and not the form $m_u.u_i.m_w$. Formally speaking, we can say that we can replace only the sequence "$m_u.u_i$" in probability expressions by a small "s"; the sequence "$m_u.u_i.m_v$" must be replaced by a capital "S", i.e., by a symbol representing, not one ψ-function, but a mixture of ψ-functions. The meaning of this formal rule is that, if a measurement is not merely a repetition of preceding measurements, the assemblage created by it irrespective of its results is a mixture.

We see that the concept *physical situation*, although determining, not one system, but a statistical assemblage, does not apply to all kinds of statistical assemblages, but is restricted to special forms of such assemblages. These are the assemblages which can be characterized by one ψ-function.

The question arises whether the term *physical situation* is wider than the term *pure case*. This would be so if there were ψ-functions of such a shape that they could not be considered as eigen-functions of any physical entity. It is, however, not necessary to make such a distinction. We can always define a physical entity of such a kind that a given ψ-function is one of the eigen-functions of this entity. Mathematically speaking, this means simply that we can

§ 23. NATURE OF PROBABILITIES

always incorporate a given function in an orthogonal set, and that we can construct an operator such that the constructed set represents the solutions of the Schrödinger equation of this operator. The physical entity so defined will, in general, be none of the entities to which names have been given; it may, for instance, be something like the product of energy and position divided by the square root of the momentum. Furthermore, we do not know whether there exists a physical method of measuring such an entity, i.e., a method of producing physically a system of that kind of ψ-function. But formally speaking, any ψ-function can be considered as an eigen-function of a physical entity. We therefore can identify the two terms *pure case* and *physical situation*, or *assemblage characterized by one ψ-function*. When we distinguish, in probability expressions such as (22), § 22, and (23), § 22, a general situation s from a situation created by a measurement m_u with the result u_i, we do so only to indicate a difference of practical realization; in principle these two situations are of the same kind.

We therefore distinguish only two kinds of statistical assemblages: those which are pure cases s, or consist of systems in the same physical situation, and those which are mixtures S.[1] With this distinction we do not introduce differences concerning the application of the calculus of probability; the laws of probability are the same in both cases. What we introduce is a classification of the form in which the values of probabilities are determined. The pure case allows us to comprise the values of the probabilities into one complex function, whereas the mixture requires, in addition, the individual enumeration of the probabilities with which the individual ψ-functions occur in the assemblage. This shows a physical peculiarity of the pure case assemblage which distinguishes this assemblage from others. The peculiarity derives from the fact that *the pure case represents the utmost degree of homogeneity attainable in quantum physics;* this case may therefore be considered as defining one specific physical situation.

The present section concludes our short exposition of the mathematical methods of quantum mechanics. Summarizing, we may enumerate as the *basic principles of quantum mechanical method* the following principles:

1) The characterization of physical entities in a given context by operators, eigen-functions, and eigen-values, determined by the first Schrödinger equation (§§ 14–16).

2) The characterization of the physical situation by a function ψ which, through the rules I–III, § 17, determines the probability

---

[1] It can be shown that every mixture S can be conceived as a mixture of ψ-functions which constitute an orthogonal set. Using the considerations given above, we therefore can say that for every mixture there exists an entity x such that the mixture can be conceived as a mixture of pure cases corresponding to measurement results of the entity x. Cf. M. Born and P. Jordan, *Elementare Quantenmechanik* (Berlin, 1930), § 59.

distributions, and is inversely determined by these distributions (§ 20).

3) The law of the time dependence of ψ, given in the second Schrödinger equation (§ 18).

4) The quantum mechanical definition of measurement (§ 21).

The principle of indeterminacy and the disturbance of the object by the measurement are not included in these basic principles; they constitute *derivable principles*, since they are derivable from the basic principles.

Part III

INTERPRETATIONS

§24. Comparison of Classical and Quantum Mechanical Statistics

In Part II we have explained the mathematical methods by which quantum mechanics derives the probability distributions of physical entities. We now shall turn to a logical analysis of these methods.

In classical physics a situation is determined if the parameters $q_1 \ldots q_n$ and $p_1 \ldots p_n$ are given. Let us write, as before, q for $q_1 \ldots q_n$, and p for $p_1 \ldots p_n$. Values at the time t_o are indicated by the subscript o. We then can summarize the *derivative relations of classical physics* in the following table:

Given:
1) the values of q_o and p_o at a time t_o
2) the physical laws

Determined:
3) the value u_o of every other entity concerned at the time t_o
4) the values q_t and p_t of q and p at a time t
5) the value u_t of any entity u at a time t

The derivation of the values 3–5 is achieved by means of the following mathematical functions, in which we use the symbol f to indicate a function in general, without specification of its individual form:

$$q_t = f_t(q_o, p_o) \qquad (1)$$

$$p_t = f_t(q_o, p_o) \qquad (2)$$

$$u_t = f(q_t, p_t) \qquad (3)$$

These relations constitute the causal laws of the problem.

Now let us consider a generalization which may be called classical-statistical physics, and which consists in a combination of causal laws with statistical methods. In this case we consider the causal laws (1)–(3) as still valid; we introduce, however, a method which takes into account the fact that the values q_o and p_o cannot be exactly measured. The specification of these values is therefore replaced by the statement of their probability distributions. We write these distributions at the time t_o in the form $d_o(q)$, $d_o(p)$, and $d_o(q,p)$, the latter function giving the probability of combinations of values q and p. The table of *derivative relations of classical-statistical physics* then can be written as follows:

Given:
1) the probability distribution $d_o(q,p)$ at the time t_o
2) the physical laws

Determined:
3) the probability distributions $d_o(q)$, $d_o(p)$, and the probability distribution $d_o(u)$ of every other entity concerned, at the time t_o
4) the probability distribution $d_t(q,p)$ at the time t
5) The probability distributions $d_t(q)$, $d_t(p)$, and the probability distribution $d_t(u)$ of every other entity concerned, at the time t

Comparing this table with that of the strictly causal case, we see that the occurrence of probability distributions involves a further distinction which originates from the distinction between $d_o(q,p)$ on the one side, and $d_o(q)$ and $d_o(p)$ on the other side: The first distribution appears in the category of what must be given, the latter two distributions appear in the category of what is determined. The derivation of the latter two distributions is achieved through the relations:

$$d_o(q) = \int d_o(q,p) dp \qquad (4)$$

$$d_o(p) = \int d_o(q,p) dq \qquad (5)$$

It is important to realize that this determination cannot be reversed, i.e., that the distribution $d_o(q,p)$ of the combinations is not determined merely by the individual distributions. Only in case we know that q and p represent independent entities can we write

$$d_o(q,p) = d_o(q) \cdot d_o(p) \qquad (6)$$

But the application of the special theorem of multiplication, expressed in (6), presupposes some definite physical knowledge about the independence of q and p. Without such additional knowledge the function $d_o(q,p)$ must be directly given; and in general it will not have the special form (6). On the other hand, if neither such additional knowledge nor a statement of the function $d_o(q,p)$ is available, the knowledge of the distributions $d_o(q)$ and $d_o(p)$ alone will not be sufficient to determine the corresponding functions $d_t(q)$ and $d_t(p)$ at a later time t.

The function $d_o(q,p)$, therefore, may be said to determine the *probability state* of the system, or its *physical situation*, so far as it can be stated in statistical terms; for, if $d_o(q,p)$ is known, all the other probability distributions are determined. As to the distributions $d_o(q)$ and $d_o(p)$, this determination is expressed in (4) and (5); the determination of the other distributions included

§24. CLASSICAL AND QUANTUM STATISTICS

in 3–5 is based on the following ideas. With $d_o(q,p)$ we know the probability of every combination of values q and p. Now the causal law (3) coordinates to every such combination a value u; the probability distribution $d_o(u)$ is therefore determined. Similarly, the causal laws (1) and (2) coordinate to every combination of values q and p a value of q, or p, at a later time; the probability distributions of these values are therefore determined.

We now shall show the way in which these derivations are carried through. Since q and p stand for sets $q_1 \ldots q_n$ and $p_1 \ldots p_n$, the function $d_o(q,p)$ stands for[1]

$$d(q_1^o \ldots q_n^o, p_1^o \ldots p_n^o) \tag{7}$$

In order to simplify our notation we shall omit the subscript t, and use the symbols $q_1 \ldots q_n$, $p_1 \ldots p_n$ for the values of these variables at the time t. Furthermore, the general symbol f, standing for function, may be specified individually; we then can write for (1):

$$q_1 = q_1(q_1^o \ldots q_n^o, p_1^o \ldots p_n^o)$$
$$q_2 = q_2(q_1^o \ldots q_n^o, p_1^o \ldots p_n^o) \tag{8}$$
$$\ldots\ldots\ldots\ldots\ldots\ldots$$
$$p_n = p_n(q_1^o \ldots q_n^o, p_1^o \ldots p_n^o)$$

Solving (8) for the variables $q_1^o \ldots p_n^o$, we can introduce the variables $q_1 \ldots p_n$ in (7). Since a probability distribution involves integrations in its variables, we must use here the rules for the introduction of new variables in integrals; we therefore must multiply with the functional determinant

$$\Delta = \begin{vmatrix} \dfrac{\partial q_1^o}{\partial q_1} & \ldots\ldots\ldots & \dfrac{\partial p_n^o}{\partial q_1} \\ \ldots\ldots\ldots\ldots\ldots\ldots \\ \ldots\ldots\ldots\ldots\ldots\ldots \\ \dfrac{\partial q_1^o}{\partial p_n} & \ldots\ldots\ldots & \dfrac{\partial p_n^o}{\partial p_n} \end{vmatrix} \tag{9}$$

and we have

$$d_t(q,p) = d(q_1 \ldots q_n, p_1 \ldots p_n) = d(q_1^o \ldots q_n^o, p_1^o \ldots p_n^o) \cdot |\Delta| \tag{10}$$

For canonical parameters q and p, such as position and momentum, Hamilton's equations lead to the result that $\Delta = 1$; this is the theorem of Liouville.

In a similar way $d_t(u)$ is determined. When we omit the subscript t, (3) can be written:

$$u = f(q_1 \ldots q_n, p_1 \ldots p_n) \tag{11}$$

---

[1] We use here a way of writing in which the symbols q_i^o, p_i^o denote, not constants, but variables. The values of q and p at the time t_o then are considered as forming a set of variables, different from the set of variables constituted by the values of q and p at a later time. This way of writing is preferable for the present purpose.

114   PART III. INTERPRETATIONS

Starting from the function $d(q_1 \ldots q_n, p_1 \ldots p_n)$, we can now substitute for one of its variables, for instance for q_1, the variable u; for this purpose we solve (11) for q_1 and put the obtained expression in the place of q_1. The resulting function may be written $d'(u,q_2 \ldots q_n, p_1 \ldots p_n)$. This is not yet a probability distribution; in order to make it so we must multiply it by the functional determinant which reduces here to the value $\frac{\partial q_1}{\partial u}$. We then obtain the probability distribution

$$d(u,q_2 \ldots q_n, p_1 \ldots p_n) = d'(u,q_2 \ldots q_n, p_1 \ldots p_n) \cdot \left|\frac{\partial q_1}{\partial u}\right| \qquad (12)$$

When we now reintroduce the subscript t and write $d_t(u)$ for $d(u)$, the function $d_t(u)$ is obtained by integrating $(n-1)$ times:

$$d_t(u) = \int \ldots n-1 \ldots \int d(u,q_2 \ldots q_n, p_1 \ldots p_n) dq_2 \ldots dq_n dp_1 \ldots dp_n \qquad (13)$$

In order to state our results symbolically, we shall use the symbol f_{op} to express an operator, without specification of the individual form of the operator, which is different in each of the following equations. We then can write:

$$d_t(q,p) = f_{op}{}^{(t)}[d_o(q,p)] \qquad (14)$$

$$d_t(q) \quad = f_{op}[d_t(q,p)] \quad = f_{op}{}^{(t)}[d_o(q,p)] \qquad (15)$$

$$d_t(p) \quad = f_{op}[d_t(q,p)] \quad = f_{op}{}^{(t)}[d_o(q,p)] \qquad (16)$$

$$d_t(u) \quad = f_{op}[d_t(q,p)] \quad = f_{op}{}^{(t)}[d_o(q,p)] \qquad (17)$$

The meaning of the operators f_{op} is given by the procedures explained above, which include the application of the causal laws (1)–(3).

Whereas (14) has no equivalent in the strictly causal case, we see that the relations between the first and the last term in (15)–(17) represent direct generalizations of the causal case (1)–(3). The only formal difference is that the values p, q, u are replaced by the functions $d(p)$, $d(q)$, $d(u)$, or $d(q,p)$, and that the function f is replaced by the operator f_{op}. This means that in the statistical case, relations between statistical functions take the place of the relations between variables of the classical case.

Now let us consider quantum mechanics. The analogue of the probability distribution $d_o(q,p)$ is here the complex function $\psi_o(q)$, if we write this term for $\psi(q,t_o)$, i.e., the form of the ψ-function at the time t_o. Like $d_o(q,p)$, $\psi_o(q)$ determines the *probability state* of the system, and with this it determines the *physical situation* of the system as completely as it is possible in quantum

§24. CLASSICAL AND QUANTUM STATISTICS

physics. We therefore can summarize the *derivative relations of quantum mechanics* in the following table:

Given:
1) the complex function $\psi_o(q)$ at the time t_o
2) the physical laws

Determined:
3) the probability distributions $d_o(q)$, $d_o(p)$, and the probability distribution $d_o(u)$ of every other entity concerned, at the time t_o
4) the complex function $\psi_t(q)$ at the time t
5) the probability distributions $d_t(q)$, $d_t(p)$, and the probability distribution $d_t(u)$ of every other entity concerned, at the time t

The physical laws mentioned in 2 include, with respect to points 3 and 5, the rules for the construction of operators and of the first Schrödinger equation, and, in addition, the rules I–III, § 17, determining the probability distributions; with respect to point 4 we include the second Schrödinger equation in the physical laws.

Using the operator form of writing applied in (14)–(17), we can write the derivative relations of quantum mechanics in the form:

$$\psi_t(q) = f_{op}{}^{(t)}[\psi_o(q)] \tag{18}$$

$$d_t(q) = f_{op}[\psi_t(q)] = f_{op}{}^{(t)}[\psi_o(q)] \tag{19}$$

$$d_t(p) = f_{op}[\psi_t(q)] = f_{op}{}^{(t)}[\psi_o(q)] \tag{20}$$

$$d_t(u) = f_{op}[\psi_t(q)] = f_{op}{}^{(t)}[\psi_o(q)] \tag{21}$$

These relations show the analogy with the classical statistical case (14)–(17). The function $d(q,p)$ is here replaced by $\psi(q)$. The role which the causal laws (1) and (2) play with respect to the establishment of (14) is assumed by the time-dependent Schrödinger equation with respect to (18).

As in the classical statistical case, the two individual distributions $d_o(q)$ and $d_o(p)$ are not sufficient to determine the probability state, since the latter distributions do not determine the function $\psi_o(q)$ (cf. § 20). The function $\psi_o(q)$ therefore resembles the function $d_o(q,p)$ so far as it states more than the individual distributions. On the other hand, we notice here a remarkable difference. The complex function $\psi_o(q)$ is determined by a pair of two other probability distributions, for instance, by the pair of two consecutive functions $d_o(q)$ and $d_1(q)$ (cf. § 20). Whereas, therefore, the complex function $\psi_o(q)$ represents a pair of two probability distributions in one variable, the function $d_o(q,p)$ is, in general, superior to such a pair. Apart from special cases such as (6), it is impossible to determine a function of two variables in terms of two functions of one variable. The same holds when we replace q and p by the sets $q_1 \ldots q_n$ and $p_1 \ldots p_n$; the function $d_o(q_1 \ldots q_n, p_1 \ldots p_n)$, which has $2n$ variables, is, in general, superior to a pair of two functions of n variables each. We see that the probability state which we call a quantum mechanical situation, or a pure case, is of a special kind; if the number of parameters of the classical problem is $2n$, the probability state of the quantum mechanical problem is expressible in terms of two probability distributions of n variables each. It is this fundamental fact which is expressed in the use of a complex function ψ for the characterization of a quantum mechanical situation. Only for a mixed case, i.e., for an assemblage not describable in terms of one ψ-function, do we need a function of $2n$ arguments comparable to $d_o(q_1 \ldots q_n, p_1 \ldots p_n)$.

The physical situations of quantum mechanics represent, therefore, statistical cases of a special type which may be compared with classical statistical cases of the independence type (6), i.e., cases in which the function $d_o(q,p)$ splits into two functions $d_o(q)$ and $d_o(p)$. This is, however, not more than an analogy. The quantum mechanical situation cannot be conceived as a situation in which the parameters q and p are independent, because, as we saw in § 20, $d_o(q)$ and $d_o(p)$ do not determine $\psi_o(q)$. Such a statement must even be considered as being beyond the reach of quantum mechanics. There is no way of deriving any statement concerning a distribution $d_o(q,p)$ in a quantum mechanical situation; therefore, a statement with respect to a special multiplicative form of this distribution cannot be made either. The reason that quantum mechanics does not need a function $d(q,p)$ is given in the fact that the transformations of quantum mechanics in the configuration space are holistic transformations, not point transformations (cf. § 12). A transition from the function $\psi(q)$ to the function $\sigma(p)$, which is such a holistic transformation, thus does not determine any coordination of values q to values p. Quantum mechanics therefore does not include any statements about simultaneous values of q and p.

The intervention of the function $d(q,p)$ in classical statistical considerations is due to the fact that these methods make use of causal laws which coordinate to a combination p, q, the later values of these entities, or values of an entity u, in the sense of a mathematical function. If such laws are necessary for the determination of those values, a knowledge of the probability of the combination p,q cannot be dispensed with. The fact that the quantum mechanical method does not use a function $d(q,p)$ therefore proves that this method does not include any assumption of such laws.

This can be understood in the following way. A causal law like (1) or (2) is to be defined, as explained in § 1, as the limit of probability relations of the form (14) or (15), resulting when both distributions $d(q)$ and $d(p)$, or the distribution $d(q,p)$, approach concentrated distributions. The probability distributions then degenerate to the values at the center of the concentrated distributions, and the operator f_{op} degenerates to a function f. Now the relation

§ 24. CLASSICAL AND QUANTUM STATISTICS

of indeterminacy shows that it is not possible to find physical situations in which both the distributions $d(q)$ and $d(p)$ are concentrated. It is not possible, therefore, to verify causal laws of the form (1)–(3); more precisely: It is not possible to verify such laws to any desired degree of approximation. There is a limit to this approximation drawn by the relation of indeterminacy. This is the reason that quantum mechanics does not use such laws.

Must we therefore insist that it be impermissible to speak of such laws? Such an attitude would mean the introduction of a restriction for which there is no need. We can as well assume the opposite attitude and permit the establishment of such laws in the sense of *definitions*, if it is not possible to introduce such laws as *verifiable statements*. The condition which such a liberalism would demand is that *any* law be permissible if only it establishes a connection between observable distributions corresponding to the numerical results of quantum mechanics. Instead of one law determining an entity u as a function of q and p we then have a *class of equivalent laws*. It is therefore the theory of equivalent descriptions (§ 5) which we can use in order to fill in the blank between the observable distributions.

For this reason we now must inquire into the various forms of supplementation by interpolation left open, for such a logical liberalism, by the established results of quantum mechanical method. The question of causal laws then will be turned into the question of whether there is a supplementation in which there are causal laws. The answer will depend on the meaning assumed for the term "causal law". This term usually includes two distinct conditions. First, it is required that the cause determine the effect univocally; second, it is demanded that the effect spread continuously through space, following the principle of action by contact. The latter condition is frequently expressed by the term *causal chain*. Anticipating our results, we may say here that our answer will be negative, that we shall find it impossible to interpret quantum mechanical relations completely in terms of causal chains. We then shall attempt to generalize the conception of causality by retaining only the second condition, while replacing the first by a condition of probability relations between cause and effect. In this sense we shall speak of *probability chains*; the precise definition will be given later. We shall be led to the result, however, that it is even impossible to interpolate such probability chains, by way of definitions, between the observable data of quantum mechanics. This negative result, which represents a statement about the whole class of equivalent descriptions, excludes the introduction of causality, in any sense, into the world of quantum mechanical objects, and thus establishes the *principle of anomaly* formulated in § 8.

§ 25. The Corpuscle Interpretation

Every interpretation which coordinates to a measured value q, one simultaneous value p, and vice versa, may be called a corpuscle interpretation. The coordination may be established in various ways. We shall outline one way which is distinguished by its relative simplicity and may be considered as the prototype of the corpuscle interpretation.

It is clear that the values of unobserved entities can be introduced only in the sense of *definitions*. We therefore cannot be required to prove that our statements about unobserved entities are *true*; all that is to be required is that our definitions be *admissible*. An interpretation consisting of several admissible definitions connected with each other will be called an *admissible interpretation*. Referring to our exposition in § 22, we can show that an admissible interpretation can be given in a rather simple way.

Our first definition in this interpretation contains the idea that the entity which is measured is not disturbed by the measurement; only entities noncommutative with it, according to this idea, are disturbed. We formulate this idea as follows:

Definition 1. If the value u_i of an entity u has been observed in a measurement of u, this value u_i means the value of u immediately before and immediately after the measurement.

If u is a time-independent entity, it follows that the same value u_i is valid also for a longer time before and after the measurement. Only an intervention from outside, such as the measurement of a noncommutative entity v, then will destroy the value u_i.

Using the terminology of § 22, we can express definition 1 in terms of probability relations as follows:

$$P(s,u_i) = P(s.m_u,u_i) \qquad (1)$$

This relation states that the value u_i exists even before the measurement m_u. That u_i exists also after the measurement m_u need not be specially expressed, because our way of writing expresses this idea by the rule that the terms follow each other in the order of time.

If a measurement of u has been made, we can express our convention with respect to a measurement of v:

$$P(m_u.u_i,v_k) = P(m_u.u_i.m_v,v_k) \qquad (2)$$

Formally we can express the relations (1) and (2) by the rule: The term m_v can be dropped in the first place of a probability expression if it is immediately followed in the second place by a term v_k.

We now shall show that our definition leads to a determination of simultaneous values of noncommutative entities. Imagine we first make a measure-

§ 25. THE CORPUSCLE INTERPRETATION

ment of u, obtaining u_i, and then a measurement of the canonically conjugated entity v, obtaining v_k. Let both entities be time independent. Since, according to our convention, the value u_i exists after the measurement of u, and the value v_k before the measurement of v, we now know two values, u_i and v_k, existing at the same time, namely, between the measurements m_u and m_v. This knowledge can be symbolized by writing

$$u_i \quad m_u \quad u_i \quad v_k \quad m_v \quad v_k \tag{3}$$

in which diagram time runs from left to right. We see that our definition makes it meaningful to speak of the combination of the canonically conjugated values u_i and v_k, existing simultaneously, since u_i as well as v_k exist throughout the time from m_u to m_v.

If we want to extend this method of determination to time-dependent entities, we must make the two measurements m_u and m_v immediately after each other; we then shall know the simultaneous values u_i and v_k between these two measurements. If only v, but not u, is time dependent, the two measurements, with respect to time, need not be close together; but the results then determine only the combination of simultaneous values existing immediately before m_v. Such cases occur, for instance, when u means the velocity of a free particle not subject to forces, and v its position.

The knowledge of simultaneous values attained by definition 1 is of a rather restricted kind, since we acquire the knowledge of a combination u_i, v_k, only after one of these values has been destroyed. Once the measurement m_v has been made, the value u_i exists no longer; we would have to make another measurement of u in order to know the then-existing value of u. But with this measurement of u the value v_k would be destroyed, etc. This result has an important consequence. In § 21 we stated the principle that repetition of a measurement yields the same value as the preceding measurement. This principle does not apply to a combination of simultaneous values. We have no means of producing the same combination of values u_i and v_k a second time; only by chance can it happen that the same combination results once more. We therefore cannot use the knowledge of simultaneous values for the prediction of the results of future measurements; the future remains undetermined, since the knowledge acquired is no longer valid for the state of affairs existing at the time of its acquisition. We may add the remark that we cannot use our knowledge for an inference into the past either, since we cannot tell from the observed combination u_i and v_k which value of v existed before the measurement m_u.[1]

Definition 1 determines values of unobserved entities only between two

[1] We may, of course, know this value from a previous observation whose result is reported and therefore known at a later time, whereas the results of future measurements cannot be known at earlier times. This shows the existence of a direction of time. But the use of reports involves manipulations which cannot be performed on a closed system. So far as closed systems are concerned, it seems that quantum mechanics does not distinguish a direction of time. This follows from (14), § 22.

measurements. In order to know something about such values in other situations, we must introduce a second definition. It is sufficient for our purposes if we define for the general case, not the value of the unobserved entity, but the probability of such a value. We shall use here the special rule of multiplication, thus constructing our interpretation by analogy with the form (6), § 24, of the classical-statistical case. This is done by the following definition:

Definition 2. The probability of a combination $v_k w_m$ relative to a physical situation s is given by the product of the individual probabilities of v_k and w_m relative to s. In symbols:
$$P(s, v_k.w_m) = P(s, v_k) \cdot P(s, w_m) \tag{4}$$

Combining this relation with (1) we have
$$P(s, v_k.w_m) = P(s.m_v, v_k) \cdot P(s.m_w, w_m) \tag{5}$$

Definition 2 is assumed to hold also in case ψ is replaced by an eigen-function of u, i.e., if ψ characterizes a situation resulting from a measurement of u. We then have
$$P(m_u.u_i, v_k.w_m) = P(m_u.u_i, v_k) \cdot P(m_u.u_i, w_m) \tag{6}$$
$$= P(m_u.u_i.m_v, v_k) \cdot P(m_u.u_i.m_w, w_m) \tag{7}$$

Definition 2 states the independence of the values v_k and w_m. Instead of expressing this idea by means of the special rule of multiplication (4), we can also express it as a statement that the relative probability of w_m with respect to v_k is equal to the "absolute" probability of w_m. This means we can replace (4) and (6), respectively, by the formulae:[2]
$$P(s.v_k, w_m) = P(s, w_m) \tag{8}$$
$$P(m_u.u_i.v_k, w_m) = P(m_u.u_i, w_m) \tag{9}$$

Applying here (1) and (2), first on the left hand sides, then on the right hand sides, we obtain:
$$P(s.v_k, w_m) = P(s.v_k.m_w, w_m) = P(s.m_w, w_m) \tag{10}$$
$$P(m_u.u_i.v_k, w_m) = P(m_u.u_i.v_k.m_w, w_m) = P(m_u.u_i.m_w, w_m) \tag{11}$$

In the substitution of (1) on the left hand side of (8) we have considered v_k as belonging to the state s, and have therefore placed the term m_w to the right of v_k. The same holds for the application of (2) to the left hand side of (9). We can also consider expressions of the form:
$$P(s.m_w.v_k, w_m) \tag{12}$$

---

[2] (8) is derivable from (4) only for the case $P(s, v_k) \neq 0$, whereas inversely (4) is derivable from (8) also for this case. We shall therefore consider (8) and (9) as the definition of independence, and thus as replacing definition 2, whenever the case $P(s, v_k) = 0$ is to be included.

§25. THE CORPUSCLE INTERPRETATION   121

Such an expression cannot be derived from $P(s.v_k,w_m)$ by means of (1). We can, however, determine the value of (12) by the following method, using, besides (1), the general rule of multiplication of probabilities, the rule of elimination corresponding to (4), § 22, and (22), § 22:

$$P(s.m_w.v_k,w_m) = \frac{P(s.m_w,v_k.w_m)}{P(s.m_w,v_k)} = \frac{P(s.m_w,w_m) \cdot P(s.m_w.w_m,v_k)}{P(s.m_w.m_v,v_k)}$$

$$= \frac{P(s.m_w,w_m) \cdot P(m_w.w_m.m_v,v_k)}{\sum_i P(s.m_w,w_i) \cdot P(m_w.w_i.m_v,v_k)} \qquad (13)$$

The relations (8) and (9) can be expressed by the following rule, which is so formulated that it does not apply to terms like (12): a term like v_k may be dropped in the first place of a probability expression if it is neither preceded by the term m_v, nor by a term m_w which is followed by w_m in the second place.

The admissibility of definition 2 is shown by the following consideration which proves that the rule of elimination is tautologically satisfied:

$$P(s,w_m) = P(s,[v_1 \vee v_2 \vee \ldots].w_m)$$

$$= \sum_k P(s,v_k.w_m)$$

$$= \sum_k P(s,v_k) \cdot P(s.v_k,w_m)$$

$$= P(s,w_m) \cdot \sum_k P(s,v_k) = P(s,w_m) \qquad (14)$$

For the last line we have used (8), and the relation $\sum_k P(s,v_k) = 1$. The same can be shown for expressions of the form (13).

Furthermore, we can show that our rule (9) does not contradict our previous rules for relative probabilities. The relation (9) states that the value w_m is independent of an *unmeasured* value v_k. The relative probabilities previously introduced, such as (15), § 22, or (22), § 22, state a dependence of w_m on a *measured* value v_k.

With the definitions 1 and 2 we have given an interpretation which is exhaustive, i.e., in which statements about unobserved entities, including simultaneous values, can be made. Using a terminology introduced in the general theory of meaning,[3] we can say that in this interpretation statements about simultaneous values of noncommutative entities have *truth meaning* for situations between two measurements, and *probability meaning* for general situations. This distinction expresses the fact that for the latter case we have defined only the probability of the values of the entities (definition 2), whereas in the first case the values themselves are defined (definition 1).

[3] Cf. the author's *Experience and Prediction* (Chicago, 1938), §7.

§ 26. The Impossibility of a Chain Structure

After the corpuscle interpretation has been established as an admissible interpretation, we can ask the question of the consequences of this interpretation. It is, in particular, the question of causality which interests us. We saw that in the classical-statistical conception there are causal laws which connect the states q,p with future states, and with other entities u. Are there similar laws introduced with our corpuscle interpretation?

A short analysis shows that this is not the case. Let us assume a sequence of values corresponding to (3), § 25, in the form

$$q_i \quad m_q \quad q_i \quad p_k \quad m_p \quad p_k \qquad (1)$$

and let u be an entity which is neither commutative with q nor with p. We then have no way of measuring u between the measurements m_q and m_p without disturbing the situation. We can only introduce a probability

$$P(m_q.q_i.p_k,u_m)$$

According to (9), § 25, this probability is equal to

$$P(m_q.q_i.p_k,u_m) = P(m_q.q_i,u_m) = P(m_q.q_i.m_u,u_m) \qquad (2)$$

This means that u does not depend on the value of p existing simultaneously with q_i, and that, on the other hand, the dependence of u on q is so constructed that to one value q there is coordinated a spectrum of possible values u. We therefore cannot introduce a causal function

$$u = f(q,p) \qquad (3)$$

such that the statistical relations of quantum mechanics are reduced to the classical-statistical case.

The latter case is constructed in such a way that it possesses *causal chains* which establish the connection between $d(q)$ and $d(p)$ on the one hand and $d(u)$ on the other hand. Our considerations therefore show that such causal chains cannot be constructed in the interpretation of definitions 1 and 2. Generalizing our investigation in the direction indicated at the end of § 24, we shall ask whether in the interpretation given by these definitions at least *probability chains* can be constructed. The meaning of this term must now be defined precisely.

For this purpose let us consider a structure in which a *relative probability function*

$$d(q,p;u) \qquad (4)$$

takes the place of the causal function (3) and establishes the connection between the distributions $d(q)$ and $d(p)$ on the one hand, and $d(u)$ on the other hand. Relative probability functions correspond to relative probabilities;

§26. CHAIN STRUCTURE

they are integrated only over the variable following the semicolon and determine the probability of this variable relative to values of the variables before the semicolon. These functions are a generalization of causal laws into probability laws.[1] We shall say that the interpretation so introduced is given in terms of *probability chains* if the function (4) is independent of the functions $d(q)$ and $d(p)$. The term *chain structure* therefore characterizes a structure in which the probability of a value u is determined if the values q and p are given. We can interpret this feature as meaning that the connection between $d(q)$ and $d(p)$ on the one side, and $d(u)$ on the other side, *goes through* the values q and p.

Our results show that we cannot even construct probability chains in the sense defined. Because of (2), the function (4) degenerates into a function

$$d(q;u) \qquad (5)$$

of q alone. Whereas, in this case, the value of p has no influence on u, it is different after a measurement of p has been made; we then have a probability function

$$d(p;u) \qquad (6)$$

which shows u to be independent of q. This means that the probability of a value u, in the given interpretation, depends sometimes on q alone, and sometimes on p alone. In the first case, therefore, the value of p, which we assumed as existing, though not being measured, has no influence on u; whereas in the second case the value of q has no influence on u. Now the two cases differ with respect to the probability distributions $d(q)$ and $d(p)$, since in the first case $d(q)$ is a concentrated distribution, and $d(p)$ is not, whereas in the second case the converse holds. We thus find that the dependence of u on q or p varies with the distributions $d(q)$ and $d(p)$; we therefore do not have a function (4) which is independent of $d(q)$ and $d(p)$. In consideration of the requirement stated above, the existing relations, therefore, cannot be interpreted as a chain structure.

This result is made clearer if we start from a general situation s. We then have, according to definition 2,

$$d(q,p;u) = d_s(u) \qquad (7)$$

where $d_s(u)$ is the distribution of u relative to the situation s. This means that the probability of a value u does not depend on the particular combination of values q,p, nor on either of these values. Instead, it depends directly on s, and with this on $d(q)$ and $d(p)$. The fact that these latter distributions, although they determine the situation s to a great extent, do not completely determine s (cf. § 20), has no influence on our result; it proves only that, besides $d(q)$ and $d(p)$, there are further factors determining s. We therefore do not have a chain structure in the sense defined.

---
[1] Cf. the author's *Wahrscheinlichkeitslehre* (Leiden, 1935), § 44.

The question arises whether this negative result is confined to the interpretation of definitions 1 and 2. We shall show that this is not the case, and that also with other definitions of the values of unobserved entities we cannot introduce a chain structure. In order to show this we shall proceed by steps, choosing more and more general forms of definitions.

We begin with the following set of definitions. We leave definition 1 unchanged, but retain definition 2 only for q and p, whereas other entities u are considered as being dependent on p and q even when u is not observed. We ask whether this dependence of u on p and q can be so constructed that a function (4) exists. If this interpretation is to represent a chain structure, we must demand, as explained above, that the same function (4) holds for all possible physical situations s. Now it can be shown that it is impossible to define a function (4) of the required property. This is proved by the following considerations.

Let $d_o(p)$ be a distribution compatible with a rather exact measurement of q; thus, $d_o(p)$ will represent a flat curve. We divide the axis q into small intervals Δq_i corresponding to the greatest exactness with which q can be measured when p has the distribution $d_o(p)$; let q_i be the mean value of an interval Δq_i. Furthermore, let $d_i(q)$ be a distribution which is practically entirely within the interval Δq_i; when we have such a distribution we can say with practical certainty that the value of q is q_i. A ψ-function which unites the two distributions $d_o(p)$ and $d_i(q)$ may be written $\psi_i(q)$. Since our definition leaves $d_i(q)$ open within the limits of the interval Δq_i (and besides for the reasons explained in § 20), there is, not one, but a class of functions $\psi_i(q)$ satisfying this condition. Let us assume that a rule which determines one ψ-function of this class for every q_i has been introduced; this function may be written $\psi(q_i,q)$. The function $d_i(q) = |\psi(q_i,q)|^2$ then is approximately given by a Dirac function. The class of situations characterized by the functions $\psi(q_i,q)$, for any interval Δq_i, may be called the class A.

If, in a situation of the class A, we measure an entity u not commutative with q, the probability of obtaining a value u is determined by the coefficients σ of the expansion

$$\psi(q_i,q) = \int \sigma(q_i,u)\varphi(u,q)du \tag{8}$$

Since $\psi(q_i,q)$ represents a measurement of q with the result q_i, we can write the probability under consideration in the form

$$P(m_q.q_i.m_u,u) = |\sigma(q_i,u)|^2 \tag{9}$$

Applying definition 1 in the form (1), § 25, we can drop the term m_u on the left hand side, and thus arrive at

$$P(m_q.q_i,u) = |\sigma(q_i,u)|^2 \tag{10}$$

§26. CHAIN STRUCTURE

Using the d-symbol, we can write the left hand side in the form $d_i(q_i;u)$, and therefore have for all systems of the class A:

$$d_i(q_i;u) = |\sigma(q_i,u)|^2 \tag{11}$$

Now let us consider a class B of systems which all possess the distribution $d_o(p)$ previously used, but which have any of the distributions $d(q)$ compatible with $d_o(p)$. This class will include the systems of class A, but will furthermore include systems with rather flat distributions $d(q)$. For each system of class B we have, according to the rule of elimination holding for probabilities, the relation:

$$d_B(q;u) = \int d_B(q;p) \cdot d(q,p;u)dp \tag{12}$$

Here we have given no subscript B to the second term under the integral because this function, introduced in (4), is supposed to be the same for all situations. Now applying definition 2, § 25, to q and p, and using the definition of class B, we have

$$d_B(q;p) = d_B(p) = d_o(p) \tag{13}$$

We therefore infer

$$d_B(q;u) = \int d_o(p) \cdot d(q,p;u)dp \tag{14}$$

This means that the distribution $d_B(q;u)$ is the same for all systems of the class B, since the right hand side of (14) has the same value for all systems of B. We can now determine the function $d_B(q;u)$ as follows. Since the systems of the class A belong to the class B, we have, using any of the intervals Δq_i:

$$d_B(q_i;u) = d_i(q_i;u) = |\sigma(q_i,u)|^2 \tag{15}$$

Since this relation holds for all i, we can omit the subscript i from q_i, and thus arrive at
$$d_B(q;u) = |\sigma(q,u)|^2 \tag{16}$$

Now the meaning of $d_B(q;u)$ is given by

$$P(s_B.q,u) = d_B(q;u) \tag{17}$$

where s_B represents any situation of the class B. Using (9) we derive now from (16) and (17) the relation

$$P(s_B.q,u) = P(m_q.q.m_u,u) \tag{18}$$

Applying (1), § 25, to the left hand side and (22), § 22, to the right hand side, we can write this in the form:

$$P(s_B.q,u) = P(s_B.q.m_u,u) = P(s_B.m_q.q.m_u,u) = P(s_B.m_q.q,u) \tag{19}$$

when we apply (1), § 25, once more in the last step. Using considerations such as presented in § 22, we can show that this result leads to contradictions when we use the rule of elimination. When we apply the latter rule formally to the term on the left hand side in each of the following relations, we have

$$P(s_B, u) = \int P(s_B, q) \cdot P(s_B.q, u) dq \tag{20a}$$

$$P(s_B.m_q, u) = \int P(s_B.m_q, q) \cdot P(s_B.m_q.q, u) dq \tag{20b}$$

Now the first terms under the integrals are equal because of (1), § 25; the equality of the second terms is stated in (19). We therefore derive, using (1), § 25, for the second line:

$$P(s_B, u) = P(s_B.m_q, u) \tag{21a}$$

$$P(s_B.m_u, u) = P(s_B.m_q.m_u, u) \tag{21b}$$

The latter equation contradicts the inequality (18), § 22, when we replace, in the latter relation, the expression $m_u.u_i$ by s_B, the term w by u, and the term v by q. More precisely: For all those entities u for which the inequality (18), § 22, holds, (21b) will be false. It follows that there are entities u such that it is impossible to use the same function $d(q,p;u)$ for all situations of the class B.

We may try to avoid the contradiction by abandoning definition 2, § 25, which we omitted already for u, also for the entities q and p. In this case q and p are no longer independent of each other, and we have a function $d(q;p)$ such that

$$d(q;p) \neq d(p) \tag{22}$$

Since p is now dependent on q, the requirement of a chain structure must be applied to the function $d(q;p)$. This means that for all systems of the class B the function $d_B(q;p)$ is the same. But then (12) leads, as before, to the result that $d_B(q;u)$ is the same for all situations of the class B; and with this the same contradictions as before are derivable.

It may be questioned whether we are entitled to extend the postulate of a chain structure to the function $d(q;p)$. Although the p_i in this case are dependent on the q_i, one might argue that this is not to be regarded as a kind of causal dependence; i.e., this dependence should not be construed as a physical law, but as being due to accidental conditions determining the probability situation and varying from case to case. The q_i and p_i should, rather, be conceived as the independent parameters of physical occurrences in the sense that they can be arbitrarily chosen, and causal dependence in the sense of a chain structure should be assumed to hold only for other entities u with respect to the q_i and p_i. If this objection is maintained, we argue as follows. If the probability situation varies from case to case, all kinds of situations should occur; mathematically speaking, this means that the three probabilities $d(q)$, $d(p)$,

§26. CHAIN STRUCTURE

$d(q;p)$, represent the fundamental probabilities which can be assumed arbitrarily. We therefore can make the following assumption.

Assumption Γ: Among the systems of class B there is a subclass B' of systems having the same distribution $d_o(q;p)$, and for every ψ-function in B there is a number of systems having this ψ-function and belonging to B', i.e., having at the same time the distributions $d_o(p)$ and $d_o(q;p)$.

This means that we can coordinate a distribution $d_o(q;p)$ to a situation having the distributions $d(q)$ and $d_o(p)$, and that there must be systems having these three distributions. It is true that we have no means of determining observationally the latter systems, i.e., the systems of the class B'. Assumption Γ has therefore the character of a convention. But should it be impossible to carry through this convention, i.e., should the possibility of *defining* such systems be excluded, this would mean that the probability $d(q;p)$ depends on the distribution $d(q)$, since the distribution $d_o(p)$ is the same for all systems of B. In other words, this would mean that the relative probability of having a value p when a value q is given depends on the distribution of the values q. The probability that a particle at the place q has the momentum p then would depend on where the other particles are situated. This would not only contradict our assumption according to which the q and p represent the independent parameters, but would introduce a kind of dependence of the p on the q which contradicts the requirements of a chain structure. Using assumption Γ, we now write (12) for the class B', in the form:

$$d_{B'}(q;u) = \int d_{B'}(q;p) \cdot d(q,p;u)dp \qquad (23)$$

Since $d_{B'}(q;p)$ is now the same for all systems of B', namely, $= d_o(q;p)$, it follows that also $d_{B'}(q;u)$ is the same for all systems of B'. With this result the same contradictions as before are derivable, since, according to assumption Γ, B' includes a corresponding subclass A' of A, namely, a class A' of systems with functions $\psi(q_i,q)$.

We have carried through this consideration for continuous variables q and p because position and momentum are, in general, represented by variables of this kind. Since a sharp measurement of continuous variables is impossible, our result must be stated, more precisely, in the form: However small the intervals Δq are chosen, it is always possible to find an entity u for which the numerical contradictions arising from (21b) are sufficiently large. We do not want to say by our proof that the classes A, A', B, B' must remain the same for smaller Δ; this is obviously not the case. A statement, therefore, of what happens in case $\Delta = 0$ is not derivable; but such statements are in any event beyond the limits of assertability in quantum mechanics. If we were to use discrete variables q_i and p_k, such statements could be made; but in this case our proof can be given in a simpler way. If q is exactly measured as being $= q_i$, we have a function $d(q_i;p_k)$; the assumption that in a general situation s, p is independent of q, i.e., that $d_s(q_i;p_k) = d_s(p_k)$, then contradicts our definition of a chain structure, since then $d_s(q_i;p_k)$ depends on $d(q)$ and $d(p)$, namely, is different according as s represents, or does not represent, a measurement of q. If, on the other hand, we put $d_s(q_i;p_k) = d(q_i;p_k)$, we can construct, using (12) with the subscript s instead of B, the same contradictions as above. For continuous variables q and p the logical situation is different, because here sharp values do not exist, and for every measurement of q within the exactness Δ we can assume that $d(q_i;p_k) = d(p_k)$, without arriving at contradictions if only $d(p_k)$ is within the limits given by the Heisenberg relation. For this case we therefore need a special proof showing that contradictions arise for other entities u, as given above.

We now must ask whether our negative result can be evaded by the abandonment of definition 1, § 25. In order to answer this question we must enter into some general considerations concerning a definition of the values of unobserved entities.

If our system included no definition of unobserved entities, it would be easy to carry through a chain structure. We then would assume that there were unknown values q and p which were connected with the unknown value u in such a way that a function (4) existed; we even could assume that this function was degenerated into a causal function (3), § 24. Furthermore, we could give arbitrary values to the unobserved entities for each experiment in such a way that the functions (4), § 24, or (3), § 24, are satisfied. The chosen values could never be shown to be invalid if only we assume that every observation disturbs the values in an unknown way. Such an interpretation, however, cannot be called a permissible supplementation of observations because the unobserved values are not connected with the observed values by means of general rules. It therefore must be considered as a requirement of an interpolated chain structure that there be general rules which determine the unobserved values as functions of the observed values, and that these rules be the same for every situation s. The latter qualification is necessary because otherwise the disturbance by the observation would depend on the probability distributions of the measured entities; in this case, the disturbance, which in itself represents a causal occurrence, would not conform to our definition of a chain structure.

Now we know that in a situation s we can make only one observation, either of q, or of p, or of u; a further observation then will start from a new situation created by the preceding observation. The required rules therefore must be so given that they determine the unobserved value of any entity as a function of the observed value alone. Let u_o be the observed value of an entity; then we can introduce two functions $f_b(u_o)$ and $f_a(u_o)$ which determine the values *before* and *after* the measurement, so that we have

$$u_b = f_b(u_o) \qquad u_a = f_a(u_o) \tag{24}$$

Now this assumption leads to the following difficulties. When we have a situation which is definite in u_o, such as exists after a measurement m_u with the result u_o, and then make a measurement of u, the value after this measurement must be the same as before this measurement. Otherwise the value of u would change between the two measurements of u; this would be a change without cause, contradicting the principle of strict causality as well as the principle of a probability chain structure. For such a situation the two functions f_b and f_a must therefore be the same. But then these two functions cannot be different in any other situation. Since such a different situation is distinguished from the previous case only by a difference in the probability distributions, an assumption stating that the two functions f_b and f_a depend on the situation s would mean that the disturbance by the measurement depends, not only on the values of the entities, but also on the probabilities with which these values occur. This is the case we excluded above as contradicting the requirements of a chain structure.

It follows that for all situations we must have

$$f_b = f_a \qquad (25)$$

But then the generalization introduced by the function f is trivial, and does not alter our previous results. Instead of considering the observed value u_o as the value before and after the measurement, we then would regard a value $f(u_o)$ as the value before and after the measurement; and the preceding considerations leading to contradictions could then be carried through in the same way as before.

The considerations given represent a general proof that it is impossible to interpret the statistical relations of quantum mechanics in terms of a chain structure. This includes a proof that these relations cannot be interpreted in terms of causal chains, since the latter constitute a special case of probability chains. With these results the proof of the principle of anomaly, to which we referred in § 8, is given.

The demonstrations of this section are illustrated by the analysis of the interference experiment which we discussed in § 7. We saw there that we cannot introduce a probability $P(A.B_1,C)$ which is independent of the occurrences at the place B_2. This is a special case of our general theorem stating that we cannot introduce an invariant function (4); the function (4) will always depend on the whole distributions $d(q)$ and $d(p)$, i.e., on the situation s.

§ 27. The Wave Interpretation

The duality of wave and corpuscle interpretation is based on the equivalence of the function $\psi(q)$ on the one side, and pairs of suitably chosen probability functions on the other side. In the corpuscle interpretation the probability functions are considered as directly physically meaningful, whereas the function ψ appears only as a mathematical abbreviation standing for those functions. In the wave interpretation it is the function $\psi(q)$ which is considered as directly physically meaningful. This function, however, is not interpreted as a combination of two probability functions coordinated to particles, but as the description of a physical state spread out over the whole space, i.e., of the wave field. Whereas a probability function $d(q)$ determines physical reality in terms of an "or", meaning that there is a particle in the place q_1, or q_2, or etc., a wave function $\psi(q)$ determines reality in terms of an "and"; it states that there is a physical state in the place q_1, *and* in the place q_2, *and* etc. The numerical values of the function d mean probabilities; the numerical values of the function ψ mean amplitudes of the field.

This conception is introduced by a definition concerning unobserved entities differing from definition 1 with respect to the rule which establishes the values existing *before* the measurement. This is the following definition.

Definition 3. The value of an entity u measured in a situation s means the value u existing after the measurement; before the measurement, and thus in the situation s, the entity u has all its possible values simultaneously.

By "possible values" we mean values which in the situation s can be expected with a certain probability greater than 0 to be the result of a measurement. If, therefore, the eigen-values of u are discrete, only these discrete values represent possible values; furthermore, if s is given by an eigen-function $\varphi_i(q)$ of u, only one value is possible, namely, the value u_i. The latter case, however, is exceptional; there will be, in general, a spectrum of possible values which are all considered as existing simultaneously in the situation s.

The logical reason that this definition can be carried through is given by the fact that statements about unobserved entities are unverifiable; the given definition is therefore compatible with all observations, as is any rule defining *one* of the possible values as the actual one. That, moreover, the given definition leads to a conception of a field, in the technical sense of the word, has a mathematical reason grounded in the form of the time-dependent Schrödinger equation: It is the fact that the function ψ is *additive*. This means that if we have two different possible descriptions ψ_1 and ψ_2 of physical states, their sum $\psi_1 + \psi_2$ is also a possible description of a physical state. We can interpret this additivity of amplitudes as meaning the superposition of two wave fields. This fact is frequently expressed by saying that the ψ-function follows the *principle of superposition*.

If we use the corpuscle interpretation, the additivity of ψ means a rather complicated causal dependence between the physical states. Since the functions ψ_1 and ψ_2 express the probability situations holding for these individual states, the function $\psi_1 + \psi_2$ expresses the probability situation of the state given by the superposition. The latter probability situation, however, is not given by the addition of the probability functions, but constructed out of these functions

§27. THE WAVE INTERPRETATION

in a rather complicated way. If we use Feenberg's rule for the determination of the ψ-function, we can write these relations as follows:

$$\left.\begin{aligned}
\psi_1(q) &= f_{op}\left[d_1(q), \frac{\partial}{\partial t}d_1(q)\right] \\
\psi_2(q) &= f_{op}\left[d_2(q), \frac{\partial}{\partial t}d_2(q)\right] \\
\psi(q) &= \psi_1(q) + \psi_2(q) \\
d(q) &= f_{op}[\psi(q)] = f_{op}\left[d_1(q), \frac{\partial}{\partial t}d_1(q), d_2(q), \frac{\partial}{\partial t}d_2(q)\right] \\
\frac{\partial}{\partial t}d(q) &= f_{op}[\psi(q)] = f_{op}\left[d_1(q), \frac{\partial}{\partial t}d_1(q), d_2(q), \frac{\partial}{\partial t}d_2(q)\right]
\end{aligned}\right\} \quad (1)$$

The last two lines state that the two functions $d(q)$ and $\frac{\partial}{\partial t}d(q)$ of the resulting situation are determined by the corresponding functions of the individual situations. The distributions of all other entities, for instance, the distribution $d(p)$, are similarly determined. The operators f_{op} used here represent the mathematical operations explained in § 20. These relations take the place of an additivity of probabilities. As we have shown in the discussion of the interference experiment, this means that the situation resulting from such superposition is constructed in terms of causal interactions which may be of such a kind that they violate the principle of action by contact.

It is an advantage of the wave interpretation that it satisfies the principle of action by contact, since the value of the resulting amplitude in a place is determined only by the individual amplitudes in the same place. The disadvantage of the wave interpretation, on the other hand, consists in the fact that this interpretation is restricted to *interphenomena*; as soon as we are concerned with *phenomena*, we must return to the corpuscle interpretation by replacing $\psi(q)$ by probability functions. Observable entities always show a punctiform character; their spatial description is written in terms of the "or", not of the "and". This fact finds its expression in definition 3 which coordinates to the entity u *after* the measurement only one value and thus abandons the wave interpretation for this case. We see here the reason that the wave interpretation cannot reintroduce determinism in the physical world. It is true that the laws controlling the ψ-function, i.e., the Schrödinger equation, have the form of causal laws if the ψ-function is considered as representing a wave field; but the determinism so established is restricted to interphenomena, and breaks down in the realm of phenomena.[1] The indeterminism holding for phenomena finds its expression in the fact that the numerical results of observations in

---
[1] Cf. also p. 31.

general cannot be strictly predicted; such predictions are bound to be probability statements.

If we interpret ψ as a wave, we must do the same with $|\psi|^2$; we therefore must consider also the function $|\psi(q,t)|^2$ as spread out over the q-space. In the cases where $|\psi|^2$ is independent of t, i.e., for stationary systems such as (2), § 18, $|\psi|^2$ then represents a static field. It also happens, however, that $|\psi|^2$ is time dependent, and in such a way that it represents oscillations. An illustration is given by the nonstationary case of (5), § 18. The function $|\psi|^2$ is here determined as

$$|\psi(q,t)|^2 = \sum_k \sum_m \sigma_k \psi_k(q,t) \sigma_m^* \psi_m^*(q,t)$$

$$= \sum_k \sum_{m \geq k} \left\{ \sigma_k \varphi_k \sigma_m^* \varphi_m^* e^{-\frac{2\pi i}{h}(H_k - H_m)t} + \sigma_k^* \varphi_k^* \sigma_m \varphi_m e^{-\frac{2\pi i}{h}(H_m - H_k)t} \right\} \quad (2)$$

We have dropped here the argument q in φ. This can be written in the form of a Fourier expansion in real functions with real coefficients a_o, a_{km}, b_{km}, which depend only on q:

$$|\psi(q,t)|^2 = a_o(q) + \sum_k \sum_{m>k} [a_{km}(q) \cos 2\pi \nu_{km} t + b_{km}(q) \sin 2\pi \nu_{km} t]$$

$$a_{km} = a_{km} + a_{mk} \qquad b_{km} = -\frac{1}{i}(a_{km} - a_{mk}) \qquad \nu_{km} = \frac{H_k - H_m}{h}$$

$$a_o = \sum_k a_{kk} = \sum_k |\sigma_k|^2 \cdot |\varphi_k|^2 \qquad a_{km} = \sigma_k \varphi_k \sigma_m^* \varphi_m^* \quad (3)$$

The fact that we have here a double summation is irrelevant; we can easily introduce a numeration in one subscript by using one of the methods of enumerating a two-dimensional lattice of whole numbers. The duality of interpretations is now carried through as follows. Interpreting $|\psi(q,t)|^2$ as a probability function $d(q,t)$ we shall say that here the probability $d(q,t)$ performs oscillations which are given by a superposition of elementary oscillations of the frequency $\nu_{km} = \frac{H_k - H_m}{h}$. In the wave interpretation we shall speak of waves given by a superposition of such frequencies; this corresponds to Bohr's interpretation of his atom model in which the ν_{km} represent the frequencies of the waves emitted by the transition of an electron from the energy level H_k to the energy level H_m. We see that in this case waves are given, not only by the function ψ, but also the function $|\psi|^2$.

The duality of wave and corpuscle interpretation is sometimes presented as a parallel to the transformation of the state function (cf. § 19). The use of the state function $\psi(q)$ then is considered as the wave interpretation, whereas the use of a state function $\sigma(p)$, or $\sigma(H)$, is conceived as a corpuscle interpretation. This conception appears even more suggestive if the latter state function is a discrete function σ_k, since the discreteness seems to symbolize corpuscles.

§27. THE WAVE INTERPRETATION

Such a conception, however, is based on a misunderstanding of the question of interpretation. The duality, or multiplicity, of state functions has no relation to the duality of wave and corpuscle interpretation. Each state function, whether it is the function $\psi(q)$ or the function σ_k, is capable of a dual interpretation. If we consider $\psi(q)$ as a wave, and correspondingly $|\psi(q)|^2$ as a static field, or also as a wave, we conceive these functions as spread out over the whole q-space; in the corpuscle interpretation, on the contrary, we conceive $\psi(q)$ as a probability amplitude, and $|\psi(q)|^2$ as the probability of finding a particle at the place q. Likewise, we have two interpretations for the state function σ_k. Let us assume that the σ_k are the coefficients of the expansion in eigen-functions of the energy H. The corpuscle interpretation then is given by the idea that there exists only one energy state H_k, although we do not know which is this value H_k; the probability for each such state is given by $|\sigma_k|^2$. The wave interpretation is represented by the assumption that all states H_k exist simultaneously, and that the product $|\sigma_k|^2 \cdot H_k$ determines the relative amount of energy contributed to the total energy H by each state H_k. The sum

$$H = \sum_k |\sigma_k|^2 H_k \tag{4}$$

then represents this *total value*, whereas in the corpuscle interpretation it represents the *average value*.

Both cases of state functions are coupled in such a way that if we use the wave interpretation for ψ, we must also use the wave interpretation for σ_k; and if we decide for the corpuscle interpretation of ψ, we must also apply the corpuscle interpretation to σ_k. This coupling is necessary because, if ψ is a real field extended over the q-space, its components φ_k are as real as ψ, and exist simultaneously.

We see that the decision for wave or corpuscle interpretation is independent of the choice of the state function. It is true that each state function describes the physical situation completely; but *each state function is capable of both interpretations*.

Let us now consider the reason why a swarm of n particles can be considered equally as a field of waves. By a "swarm" we understand a number n of particles so far distant from each other that their mutual interaction can be neglected. In this case the energy operator H_{op} splits into a sum of operators

$$H_{op} = H_{op}^{(1)} + H_{op}^{(2)} + \ldots \tag{5}$$

such that each operator $H_{op}^{(m)}$ concerns only one particle, i.e., contains only the three coordinates of the mth particle. Let $\varphi_{k_m}^{(m)}$ be the eigen-functions of the operator $H_{op}^{(m)}$; they will then satisfy the first Schrödinger equation

$$H_{op}^{(m)} \varphi_{k_m}^{(m)} = H_{k_m} \cdot \varphi_{k_m}^{(m)} \tag{6}$$

Then the Schrödinger equation of the whole system

$$H_{op}\varphi_k = H_k \cdot \varphi_k \qquad (7)$$

is satisfied by the solution

$$\varphi_k = \varphi_{k_1 k_2 \ldots k_n} = \varphi_{k_1} \cdot \varphi_{k_2} \cdots \varphi_{k_n} \qquad H_k = H_{k_1} + H_{k_2} + \ldots H_{k_n} \qquad (8)$$

This means that the eigen-function $\varphi_{k_1 \ldots k_n}$ of the whole swarm is the product of the eigen-functions of the individual particles. The proof is easily given by putting for H_{op} in (7) its value from (5):

$$H_{op}\varphi_k = H_{op}^{(1)}\varphi_k + H_{op}^{(2)}\varphi_k + \ldots + H_{op}^{(n)}\varphi_k$$

$$= \varphi_{k_2} \cdots \varphi_{k_n} H_{op}^{(1)}\varphi_{k_1} + \varphi_{k_1}\varphi_{k_3} \cdots \varphi_{k_n} H_{op}^{(2)}\varphi_{k_2} + \ldots + \varphi_{k_1} \cdots \varphi_{k_{n-1}} H_{op}^{(n)}\varphi_{k_n}$$

$$= (H_{k_1} + \cdots + H_{k_n}) \cdot \varphi_{k_1} \cdots \varphi_{k_n} \qquad (9)$$

We can therefore consider a ψ-function of the form

$$\psi_{k_1 \ldots k_n}(q,t) = \varphi_{k_1} \cdots \varphi_{k_n} \cdot e^{-\frac{2\pi i}{h}(H_{k_1} + \ldots + H_{k_n})t} \qquad (10)$$

which splits into a product of ψ-functions of the form

$$\psi_{k_m}(q,t) = \varphi_{k_m}(q) \cdot e^{-\frac{2\pi i}{h} H_m t} \qquad (11)$$

as the definition of a physical situation which can be interpreted as a swarm of particles in a stationary state. Correspondingly, the general nonstationary state of a swarm of particles is characterized by a linear superposition of functions $\psi_{k_1 \ldots k_n}(q,t)$ of the form (11), i.e., by a ψ-function

$$\psi(q,t) = \sum_{k_1 \ldots k_n} \sigma_{k_1 \ldots k_n} \varphi_{k_1} \cdots \varphi_{k_n} \cdot e^{-\frac{2\pi i}{h}(H_{k_1} + \ldots + H_{k_n})t} \qquad (12)$$

The possibility of a corpuscular interpretation of a ψ-function of this kind, as a swarm of particles, is based on the corpuscular interpretation of the individual functions $\psi_{k_1 k_2}\ldots$ of which the total ψ-function is composed. If, on the other hand, these individual ψ-functions are interpreted as waves, the swarm obtains a wave-character. Whereas H_k in the corpuscle interpretation represents the energy of a particle, it determines in the wave interpretation the frequency through the relation $\nu_k = \dfrac{H_k}{h}$. In the corpuscle interpretation the square $|\psi(q,t)|^2$ means a probability density and thus determines the relative number of particles for a given place q at the time t; in the wave interpretation the same amount represents the intensity of the field.

Conditions similar to the kind described hold in electro-magnetic fields, although this case is combined with further mathematical complications which we cannot present in this book. Suffice it to say that the corpuscle interpretation of light waves is analogous to the duality of interpretations existing for a swarm of electrons.

§27. THE WAVE INTERPRETATION

There are two difficulties frequently mentioned in combination with the wave interpretation. One is that the values of ψ are complex numbers. This, however, cannot be considered as an obstacle to a physical interpretation. A complex number is as good for the description of physical entities as is a real number; the complex function is to be conceived as an abbreviation standing for two real functions. We know from the considerations of § 20 that these two functions represent, indirectly, two probability distributions. If we consider ψ as a wave, its complex value can be interpreted as meaning that the physical wave field is characterized, not by one scalar function, but by two. Wave fields of such a dual character are known from the electro-magnetic fields which include both an electric and a magnetic vector.

The other difficulty is more serious; it consists in the fact that the ψ-waves are functions in the configuration space of the n parameters $q_1 \ldots q_n$, whereas ordinary waves are functions in the 3-dimensional space. If we wish to interpret these n-dimensional waves as waves of the 3-dimensional space, we are led to wave fields of a very complicated structure. Let us begin with a case of $n = 6$ parameters q corresponding to the Cartesian coordinates of two particles. Translating the 6-dimensional function $\psi(q_1 \ldots q_6)$ into 3-dimensional functions, we then must start from the fact that the function $\psi(q_1 \ldots q_6)$ coordinates an amplitude, not to each point of the 3-dimensional space, but to each pair of points. This can be interpreted as meaning a set of wave fields such that there is a wave field, filling the whole space, for each point as starting point. This set of wave fields cannot be replaced by the indication of one wave field conceived as the superposition of the set. Such a superposition can be mathematically constructed by integrating $\psi(q_1 \ldots q_6)$ over three of its variables; but even if we thus construct two 3-dimensional functions, by integrating first over q_1, q_2, q_3 and then over q_4, q_5, q_6, the resulting two functions are not equivalent to the function $\psi(q_1 \ldots q_6)$. We are therefore compelled to describe the situation in 3-dimensional space as an infinite set of wave fields.

Furthermore, we must be aware of the fact that a function resulting from partial integration of a ψ-function does not have the properties of a ψ-function. This means that such a function does not describe a physical situation, and that its square does not determine a probability distribution. Thus the function

$$g(q_1,q_2,q_3) = \left| \int \psi(q_1 \ldots q_6) dq_4 dq_5 dq_6 \right|^2 \tag{13}$$

does not represent the probability of observing the particle number 1 at the place q_1, q_2, q_3. This probability is rather determined by the function

$$d(q_1,q_2,q_3) = \int |\psi(q_1 \ldots q_6)|^2 \, dq_4 dq_5 dq_6 \tag{14}$$

which is different from (13).

For a greater number of variables the 3-dimensional description is even more complicated. Thus, for three particles with nine Cartesian coordinates, we have a value ψ coordinated to every triplet of points. This means that we have a set of wave fields such that for every pair of two points there is a 3-dimensional wave field. In the general case of r particles we have a set such that for any combination of $(r - 1)$ points there is a 3-dimensional wave field. If the $q_1 \ldots q_n$ contain other than Cartesian coordinates, for instance, parameters of rotation, we must introduce sets of wave fields characterized by values of these parameters.

These considerations show that the ψ-waves in general do not have the character of ordinary wave fields; this is the case only when the q-space reduces to three dimensions, i.e., when single particles are concerned. Of this kind are problems of swarms of particles for which the interaction between the individual particles can be neglected. This is the reason that a wave interpretation in the usual sense of the word can be carried through for light rays and electron beams.

§ 28. Observational Language and Quantum Mechanical Language

In the preceding sections we have seen that every exhaustive interpretation leads to causal anomalies. We now shall inquire whether these causal anomalies can be avoided by the use of restrictive interpretations. The answer will turn out to be in the affirmative.

Before we can present such interpretations we must enter into closer analysis of the logical nature of interpretations, and inquire into the conditions under which the world of unobserved entities can be called complete.

Problems of epistemology are greatly simplified when we turn from the consideration of physical worlds to the consideration of physical languages.[1] Questions concerning the existence of physical entities are thus transformed into questions of the meaning of propositions. This has the great advantage that they can be discussed soberly as questions of logic outside the atmosphere of metaphysical preconceptions. For our problem a linguistic analysis can be carried through in the following form.

We have an *observational language* and a *quantum mechanical language*. The observational language contains terms such as "Geiger counter", "Wilson cloud chamber", "black line on a photographic film", "indication of a dial", etc.; the phrases "measurement of u" and "the result of the measurement of u" are defined in terms of these elementary expressions. Similarly, a physical situation s can be defined in observational terms; a way in which this is done is indicated in § 20 with the observational determination of the ψ-function.

---

[1] The importance of this method for philosophical analysis has been brought to the fore in recent years by R. Carnap, *Logical Syntax of Language* (London, 1937).

§28. OBSERVATIONAL AND QUANTUM LANGUAGE 137

The quantum mechanical language contains terms like "position q of an electron" and "momentum p of an electron". Between the two languages there exists the following relation: The truth and falsehood of statements of the quantum mechanical language is defined in terms of the truth and falsehood of statements of the observational language. We say, for instance, "the electron has the position q", when we know that the statement, "a measurement of position has been made and its result was q", is true.

This relation between *truth values* of statements in the two languages can be conceived as a relation of *meanings*. The meanings of statements of the quantum mechanical language are definable in terms of the meanings of statements of the observational language. Usually this relation is conceived as an *equivalence* of meanings. In this interpretation a quantum mechanical statement A has the same meaning as the set of observational statements $a_1 \ldots a_n$ which verify A. This is not strictly correct; the relation of meanings is more complicated because we have no absolute verification. We can only say: A is highly probable if the set $a_1 \ldots a_n$ is true. For the present purpose, however, we need not enter into a consideration of this question; we may neglect the probability factor in the correlation of A and $a_1 \ldots a_n$ and consider the meaning of A as given by the meaning of $a_1 \ldots a_n$.

It is important to realize that such a statement about meaning, with or without consideration of the probability factor, is always based on *definitions*. We *define* A in terms of $a_1 \ldots a_n$. Without such definitions a quantum mechanical language could not be established. The definitions 1 and 2, § 25, and definition 3, § 27, constitute examples of such definitions. It is therefore no objection to our exhaustive interpretation of § 25 that it uses definitions; any interpretation will do so.

Judged from the standpoint of observational predictions, no interpretation and therefore no quantum mechanical language at all is necessary; we then need not speak of electrons and their speeds and positions, but can say everything in terms of instruments of measurements. We shall say, for instance, "if a measuring instrument of a certain kind is used in such and such observational conditions, its dial will show this or that number". Only if we wish to introduce statements about microcosmic entities must we use definitions.

Now let us turn to an analysis of the observational language coordinated to quantum mechanics. As to this question, the results of § 24 lead to important consequences. We saw in § 24 that, so far as predictable statistics of individual entities are concerned, quantum mechanics is equal to classical statistics. A difference is to be found only in the way of derivation; classical statistics uses here, and must do so, an assumption about the distribution of combinations q and p, whereas quantum mechanics does not use such an assumption. Let us ask what the omission of this assumption means when we judge it in terms of the observational language.

When we say that, in a general physical situation s, simultaneous values of

138 PART III. INTERPRETATIONS

q and p cannot be ascertained, this statement of the quantum mechanical language is to be translated into observational language as follows. The operation *measurement of the entity q*, abbreviated m_q, is defined in terms of macrocosmic manipulations by the use of devices like Geiger counters, X-ray tubes, dials, etc.; in a similar way the operation m_p is defined. From these definitions we can infer by considerations remaining entirely within the observational language that the two operations m_q and m_p cannot be performed simultaneously. It is true the observational definition of a measurement is chosen with respect to quantum mechanical considerations; this is the reason that, for instance, a measurement of the momentum is defined in terms of macrocosmic devices which exclude the presence of extremely short waves. But once the observational definition is given, the incompatibility of m_q and m_p is a merely macrocosmic affair.

It follows that within the observational language we cannot raise a question with respect to what would happen if the observations m_q and m_p were made simultaneously. Such a question is as unreasonable as, for instance, the question "what would happen if the sun's temperature were at the same time one hundred degrees, and one million degrees?" The absence of statements about simultaneous values of entities does not lead, therefore, to blanks in the observational language.

Combining this result with the results of § 24 according to which the statistics of individual entities are fully determined by quantum mechanics, we arrive at the statement that the observational language of quantum mechanics is *statistically complete*. By "statistically complete" we mean that for every possible situation, defined in observational terms, the observational result of a measurement can be predicted with a determinate probability. We can express the given statement, therefore, also by saying that *the predictive methods of quantum mechanics are statistically complete in observational terms*. If we were able to know causal laws and to apply classical statistics to given probability distributions of initial conditions which satisfy the relation of uncertainty, our predictions so constructed would not be superior to quantum mechanical predictions so far as observations are concerned; every observational result that could be predicted by classical-statistical methods would be equally predictable by quantum mechanical methods.

This result has an important bearing on the question of interpretations. Whatever interpretation we give, i.e., however we construct the quantum mechanical language, it will always be statistically complete in observational terms. When we call one interpretation incomplete, or less complete than another, this word "complete" must have a meaning other than "complete in observational terms".

This is indeed the case. When we consider a statement about simultaneous values of q and p, or at least about the statistical distribution $d(q,p)$, as necessary for a complete description, the reason is given in considerations concerning the micro-world. Here the entities q and p are interpreted as being constituents of the space-time description of a physical state. If the operations m_q and m_p, observationally defined, are incompatible, we wish to have other ways of knowing simultaneous values of q and p; this desire is satisfied by suitable definitions which determine these values by other means than measurements. An interpretation which does not include such definitions is *statistically incomplete with respect to space-time description*, although it will be *statistically complete with respect to testability*.

In order to avoid the ambiguities of the word "complete", we have introduced in § 8 a terminology which distinguishes between exhaustive and restrictive interpretations. An interpretation like that of § 25, in which the values of unobserved entities are completely defined, is *exhaustive*; interpretations in which those values remain undefined are *restrictive*. The preceding results make it clear that a restrictive interpretation will always be complete in observational terms.

Restrictive interpretations are introduced with the intention of eliminating the causal anomalies ensuing for exhaustive interpretations. The results of § 26 have made it clear that such anomalies will always arise if we give a complete set of definitions like definitions 1 and 2, § 25, for the values of unobserved entities. If we want to exclude statements involving causal anomalies from the assertions of quantum mechanics, we must therefore use an incomplete set of definitions. We thus restrict the domain of quantum mechanical assertions; this is the reason that we speak of a restrictive interpretation.

The results of § 24 make it clear that restrictive interpretations will at least be *individually complete* in the statistical sense. We mean by this term that they are statistically complete with respect to statements about every individual entity; what is excluded are only statements about the combinations of values of entities, namely, noncommutative entities.

Restrictive interpretations can be given in two ways. In the first way the undesired statements are ruled out of quantum mechanical language by a definition of meaning which makes such statements *meaningless*. We shall therefore speak here of an *interpretation by a restricted meaning*. In the second way the undesired statements are excluded, not from quantum mechanical *language*, but from quantum mechanical *assertions*; this is done by ascribing to those statements a third truth value, called *indeterminate*. This interpretation may therefore be called an *interpretation by a restricted assertability*.

We now shall turn to an analysis of these two restrictive interpretations.

§ 29. Interpretation by a Restricted Meaning

The interpretation by a restricted meaning which we shall present here formulates, on the whole, ideas which have been developed by Bohr and Heisenberg. We therefore shall call this conception the *Bohr-Heisenberg interpretation*,

140   PART III. INTERPRETATIONS

without intending to say that every detail of the given interpretation would be endorsed by Bohr and Heisenberg.

This interpretation does not use our definitions 1 and 2, § 25. For the values of measured entities it uses the following definition:

Definition 4. The result of a measurement represents the value of the measured entity immediately after the measurement.

This definition contains only the common part of our definitions 1, § 25, and 3, § 27; a statement concerning the value of the entity *before* the measurement is omitted. In this interpretation we therefore can no longer say that the observed entity remains undisturbed; both the observed and the unobserved entity may be disturbed. On the other hand, such a disturbance of the measured entity is *not asserted*; this question is deliberately left unanswered by definition 4. The restrictive interpretation neither asserts nor denies a disturbance by the measurement. The interval "immediately after" of definition 4 is determined by the time dependence of the ψ-function; in a stationary state this interval may be chosen as long as we wish.

It is an immediate consequence of the restriction to definition 4 that simultaneous values cannot be measured. The considerations which we attached to definition 1 are no longer applicable, and if we measure first q and then p, the obtained values of q and p do not represent simultaneous values; only q represents a value existing between these two measurements, whereas p represents a value existing after the second measurement, when the value q is no longer valid. Only if a measurement of an entity u is followed by a measurement of the same kind, the value u obtained in the first measurement represents the value existing before and after the second measurement; but it is valid before the second measurement only because it represents, jointly, the value after the first measurement. Only a measurement following an equal measurement, therefore, is known not to disturb the entity measured.

The reason we interpret the measured value, in definition 4, as the value *after* the measurement, and not *before* the measurement, is as follows. We know that a repetition of the measurement produces the same value; therefore, if the obtained value holds for the time before the second measurement, it must also hold for the time after the first. Thus, if we replace the term "after" in definition 4 by the term "before", we are led to the consequence that the measured value holds before *and* after each measurement; this means that we have introduced definition 1. The asymmetry of time with respect to measurements, expressed in this consideration, is a characteristic feature of quantum mechanics (cf., however, fn. 1, p. 119).

We said that if q has been measured, we do not know the value of p. This lack of knowledge is considered in the Bohr-Heisenberg interpretation as making a statement about p *meaningless*. It is here, therefore, that this interpretation introduces a rule restricting quantum mechanical language. This is expressed in the following definition.

§ 29. RESTRICTED MEANING

Definition 5. In a physical state not preceded by a measurement of an entity u, any statement about a value of the entity u is meaningless.

In this definition we are using the term "statement" in a sense somewhat wider than usual, since a statement is usually defined as having meaning. Let us use the term "proposition" in this narrower sense as including meaning. Then, definition 5 states that not every statement of the form "the value of the entity is u", is a proposition, i.e., has meaning. Because it uses definition 5, the Bohr-Heisenberg interpretation can be called an *interpretation by a restricted meaning*.

We must add a remark concerning the logical form of definition 5. Modern logic distinguishes between *object language* and *metalanguage*; the first speaks about physical objects, the second about statements, which in turn are referred to objects.[1] The first part of the metalanguage, *syntax*, concerns only statements, without dealing with physical objects; this part formulates the structure of statements. The second part of the metalanguage, *semantics*, refers to both statements and physical objects. This part formulates, in particular, the rules concerning truth and meaning of statements, since these rules include a reference to physical objects. The third part of the metalanguage, *pragmatics*, includes a reference to persons who use the object language.[2]

Applying this terminology to the discussion of definition 5, we arrive at the following result: Whereas definition 4, and likewise definitions 1-2, § 25, and definition 3, § 27, determine terms of the object language, namely, terms of the form "value of the entity u", definition 5 determines a term of the metalanguage, namely, the term "meaningless". It is therefore a semantical rule. We can express it in table 1. We denote the statement "a measurement of u is made" by m_u, and the statement "the indication of the measuring instrument shows the value u" by u. These two statements belong to the observational language. Quantum mechanical statements are written with capital letters; and U expresses the statement "the value of the entity immediately after the measurement is u".[3] Table 1 shows the coordination of the two languages.

---

[1] A means of indicating the transition from object language to metalanguage is in the use of quotes; similarly, italics can be used. We shall use, for our presentation of logic, italics instead of quotes for symbols denoting propositions, in combination with the rule that symbols of operations standing between symbols of propositions are understood to apply to the propositions, not to their names (i.e., autonomous usage of operations, in the terminology of R. Carnap). Thus, we shall write "*a* is true", not " 'a' is true", and "*a.b* is true", not " '*a.b*' is true". All formulae given by us are therefore, strictly speaking, not formulae of the object language, but descriptions of such formulae. For most practical purposes, however, it is permissible to forget about this distinction.

[2] This distinction has been carried through by C. W. Morris, "Foundations of the Theory of Signs," *International Encyclopedia of Unified Science*, Vol. I, no. 2 (Chicago, 1938).

[3] More precisely: "immediately after the time for which the truth value of the statement m_u is considered". If m_u is true, this means the same as "immediately after the measurement"; and if m_u is false, we thus equally coordinate to U a time at which its stipulated truth value holds. (The expression "after the measurement" then would be inapplicable.)

142   PART III. INTERPRETATIONS

A meaningless statement is not subject to propositional operations; thus, the negation of a meaningless statement is equally meaningless. Similarly, a combination of a meaningful and a meaningless statement is not meaningful. If the statement a is meaningful, and b meaningless, then a and b is meaningless; so is a or b. Not even the assertion of the *tertium non datur*, b or $non\text{-}b$, is meaningful. The given restriction of meaning therefore cuts off a large section from the domain of quantum mechanical language. With this, all statements referring to causal anomalies are cut off (cf. pp. 40–41).

The only justification of definition 5 is that it eliminates the causal anomalies. This should be clearly kept in mind. It would be wrong to argue that statements about the value of an entity before a measurement are meaningless

TABLE 1

Observational language		Quantum mechanical language
m_u	u	U
true	true	true
true	false	false
false	true	meaningless
false	false	meaningless

because they are not verifiable. Statements about the value after the measurement are not verifiable either. If, in the interpretation under consideration, the one sort of statement are forbidden and the other admitted, this must be considered as a rule which, logically speaking, is arbitrary, and which can be judged only from the standpoint of expediency. From this standpoint its advantage consists in the fact that it eliminates causal anomalies, but that is all we can say in its favor.

It is often forgotten that the Bohr-Heisenberg interpretation uses definition 4. This definition seems so natural that its character as a definition is overlooked. But without it, the interpretation could not be given. Applying the language used in Part I we can say that definition 4 is necessary for the transition from observational data to phenomena; it defines the phenomena. It is therefore incorrect to say that the Bohr-Heisenberg interpretation uses only verifiable statements. We must say, instead, that it is an interpretation using a weaker definition concerning unobserved entities than other interpretations, and using a restricted meaning, with the advantage of thus excluding statements about causal anomalies.

We now must consider the relations between noncommutative entities. We know that if a measurement of q is made, a measurement of p cannot be

§ 29. RESTRICTED MEANING 143

made at the same time, and vice versa. Statements about simultaneous values of noncommutative entities are called *complementary statements*. With definition 5 we have therefore the following theorem:

Theorem 1. If two statements are complementary, at most one of them is meaningful; the other is meaningless.

We say "at most" because it is not necessary that one of the two statements be meaningful; in a general situation s, determined by a ψ-function which is not an eigen-function of one of the entities considered, both statements will be meaningless.

Theorem 1 represents a physical law; it is but another version of the commutation rule, or of the principle of indeterminacy, which excludes a simultaneous measurement of noncommutative entities. We see that with theorem 1 a physical law has been expressed in a semantical form; it is stated as a rule for the meaning of statements. This is unsatisfactory, since, usually, physical laws are expressed in the object language, not in the metalanguage. Moreover, the law formulated in theorem 1 concerns linguistic expressions which are not always meaningful; the law states the conditions under which these expressions are meaningful. Whereas such rules appear natural when they are introduced as conventions determining the language to be used, it seems unnatural that such a rule should assume the function of a law of physics. The law can be stated only by reference to a class of linguistic expressions which includes both meaningful and meaningless expressions; with this law, therefore, meaningless expressions are included, in a certain sense, in the language of physics.

The latter fact is also illustrated by the following consideration. Let $U(t)$ be the propositional function "the entity has the value u at the time t". Whether $U(t)$ is meaningful at a given time t depends on whether a measurement m_u is made at that time. We therefore have in this interpretation propositional functions which are meaningful for some values of the variable t, and meaningless for other values of t.

The question arises whether it is possible to construct an interpretation which avoids these disadvantages. An interesting attempt to construct such an interpretation has been made by M. Strauss.[4] Although the rules underlying this interpretation are not expressly stated, it seems that they can be construed in the following way.

Definitions 1–2, § 25, and definition 3, § 27, are rejected. Definition 4 is maintained. Instead of definition 5, the following definition is introduced.

[4] M. Strauss, "Zur Begründung der statistischen Transformationstheorie der Quantenphysik," *Ber. d. Berliner Akad., Phys.-Math. Kl.*, XXVII (1936), and "Formal Problems of Probability Theory in the Light of Quantum Mechanics," *Unity of Science Forum, Synthese* (The Hague, Holland, 1938), p. 35; (1939), pp. 49, 65. In these writings Strauss also develops a form of probability theory in which my rule of existence for probabilities is changed with respect to complementary statements. Such a change is necessary, however, only if an incomplete notation is used, such as is done by us in the beginning of § 22. In a notation which uses the term m_u in the first place of probability expressions, such as introduced in (13), § 22, a change of the rule of existence can be avoided.

144   PART III. INTERPRETATIONS

Definition 6. A quantum mechanical statement U is meaningful if it is possible to make a measurement m_u.

It follows that all quantum mechanical statements concerning individual entities are meaningful, since it is always possible to measure such an entity. Only when U stands for the combination of two complementary statements P and Q is it not possible to make the corresponding measurement; therefore a statement like P *and* Q is meaningless. Similarly, other combinations are considered meaningless, such as P *or* Q. The logic of quantum mechanics is constructed, according to Strauss, so that not all statements are *connectable;* there are *nonconnectable* statements.

It may be regarded as an advantage that this interpretation constructs the language of physics in such a way that it contains only meaningful elements. A determination concerning meaningless expressions is formulated only in terms of the rules for the connection of statements. On the other hand, the physical law of complementarity is once more expressed as a semantic rule, not as a statement of the object language.

Leaving it open whether or not the latter fact is to be considered as a disadvantage, we now must point out a serious difficulty with this interpretation resulting from definition 6. If we consider U as a function of t, it is indeed always possible to measure the entity u, and therefore U is always meaningful. It is different when we consider U as a function of the general physical situation s characterized by a general function ψ. Is it possible to measure the entity u in a general situation s? Since we know that the measurement of u destroys the situation s, this is not possible. Thus, it follows that in a general situation s even the individual statement U is meaningless. The given interpretation therefore is reduced to the interpretation based on definition 5. On the other hand, if the meaning of definition 6 is so construed that it is possible to measure u even in a general situation s, the obtained value u must mean the value of the entity in the situation s and therefore before the measurement. The interpretation is thereby shown to use definition 1, § 25. But if the latter definition is used, it is possible to make statements about simultaneous values between two measurements, such as explained in § 25; this means that the rules concerning nonconnectable statements break down.

If we are right, therefore, in considering Strauss's interpretation as given through definition 4 and definition 6, we come to the result that this interpretation leads back to the interpretation of definition 4 and definition 5.[5]

§ 30. Interpretation through a Three-valued Logic

The considerations of the preceding section have shown that, if we regard statements about values of unobserved entities as meaningless, we must include meaningless statements of this kind in the language of physics. If we

[5] Mr. Strauss informs me that he is planning to publish a new and somewhat modified presentation of his conceptions.

§ 30. THREE-VALUED LOGIC

wish to avoid this consequence, we must use an interpretation which excludes such statements, not from the domain of *meaning*, but from the domain of *assertability*. We thus are led to a three-valued logic, which has a special category for this kind of statements.

Ordinary logic is two-valued; it is constructed in terms of the truth values *truth* and *falsehood*. It is possible to introduce an intermediate truth value which may be called *indeterminacy*, and to coordinate this truth value to the group of statements which in the Bohr-Heisenberg interpretation are called *meaningless*. Several reasons can be adduced for such an interpretation. If an entity which can be measured under certain conditions cannot be measured under other conditions, it appears natural to consider its value under the latter conditions as indeterminate. It is not necessary to cross out statements about this entity from the domain of meaningful statements; all we need is a direction that such statements can be dealt with neither as true nor as false statements. This is achieved with the introduction of a third truth value of indeterminacy. The meaning of the term "indeterminate" must be carefully distinguished from the meaning of the term "unknown". The latter term applies even to two-valued statements, since the truth value of a statement of ordinary logic can be unknown; we then know, however, that the statement is either true or false. The principle of the *tertium non datur*, or of the *excluded middle*, expressed in this assertion, is one of the pillars of traditional logic. If, on the other hand, we have a third truth value of indeterminacy, the *tertium non datur* is no longer a valid formula; there is a tertium, a middle value, represented by the logical status *indeterminate*.

The quantum mechanical significance of the truth-value *indeterminacy* is made clear by the following consideration. Imagine a general physical situation s, in which we make a measurement of the entity q; in doing so we have once and forever renounced knowing what would have resulted if we had made a measurement of the entity p. It is useless to make a measurement of p in the new situation, since we know that the measurement of q has changed the situation. It is equally useless to construct another system with the same situation s as before, and to make a measurement of p in this system. Since the result of a measurement of p is determined only with a certain probability, this repetition of the measurement may produce a value different from that which we would have obtained in the first case. The probability character of quantum mechanical predictions entails an absolutism of the individual case; it makes the individual occurrence unrepeatable, irretrievable. We express this fact by regarding the unobserved value as indeterminate, this word being taken in the sense of a third truth value.

The case considered is different, with respect to logical structure, from a case of macrocosmic probability relations. Let us assume that John says, "If I cast the die in the next throw, I shall get 'six' ". Peter states, "If I cast the die, instead, I shall get 'five' ". Let John throw the die, and let 'four' be his result;

we then know that John's statement was false. As to Peter's statement, however, we are left without a decision. The situation resembles the quantum mechanical case so far as we have no means of determining the truth of Peter's statement by having Peter cast the die after John's throw; since his throw then starts in a new situation, its result cannot inform us about what would have happened if Peter had cast the die in John's place. Since casting a die, however, is a macrocosmic affair, we have in principle other means of testing Peter's statement after John has thrown the die, or even before either has thrown the die. We should have to measure exactly the position of the die, the status of Peter's muscles, etc., and then could foretell the result of Peter's throw with as high a probability as we wish; or let us better say, since we cannot do it, Laplace's superman could. For us the truth value of John's

TABLE 2

Observational language		Quantum mechanical language
m_u	u	U
T	T	T
T	F	F
F	T	I
F	F	I

statement will always remain unknown; but it is not *indeterminate*, since it is possible in principle to determine it, and only lack of technical abilities prevents us from so doing. It is different with the quantum mechanical example considered. After a measurement of q has been made in the general situation s, even Laplace's superman could not find out what would have happened if we had measured p. We express this fact by giving to a statement about p the logical truth-value *indeterminate*.

The introduction of the truth-value *indeterminate* in quantum mechanical language can be formally represented by table 2, which determines the truth values of quantum mechanical statements as a function of the truth values of statements of the observational language. We denote truth by T, falsehood by F, indeterminacy by I. The meaning of the symbols m_u, u, and U is the same as explained on page 141.[1]

Let us add some remarks concerning the logical position of the quantum mechanical language so constructed. When we divide the exhaustive interpretations into a *corpuscle language* and a *wave language*, the language introduced by table 2 may be considered as a *neutral language*, since it does not determine one of these interpretations. It is true that we speak of the measured entity

---

[1] As to the use of the functor, "the value of the entity," cf. fn. 4, p. 159.

§ 30. THREE-VALUED LOGIC

sometimes as the path of a particle, sometimes as the path of needle radiation; or sometimes as the energy of a particle, sometimes as the frequency of a wave. This terminology, however, is only a remainder deriving from the corpuscle or wave language. Since the values of unobserved entities are not determined, the language of table 2 leaves it open whether the measured entities belong to waves or corpuscles; we shall therefore use a neutral term and say that the measured entities represent parameters of *quantum mechanical objects*. The difference between calling such a parameter an energy or a frequency then is only a difference with respect to a factor h in the numerical value of the parameter. This ambiguity in the interpretation of unobserved entities is made possible through the use of the category *indeterminate*. Since it is indeterminate whether the unmeasured entity has the value u_1, or u_2, or etc., it is also indeterminate whether it has the values u_1, and u_2, and etc., at the same time; i.e., it is indeterminate whether the quantum mechanical object is a particle or a wave (cf. p. 130).

The name *neutral language*, however, cannot be applied to the language of table 1, page 142. This language does not include statements about unobserved entities, since it calls them meaningless; it is therefore not equivalent to the exhaustive languages, but only to a part of them. The language of table 2, on the contrary, is equivalent to these languages to their full extent; to statements about unmeasured entities of these languages it coordinates indeterminate statements.

Constructions of multivalued logics were first given, independently, by E. L. Post[2] and by J. Lucasiewicz and A. Tarski.[3] Since that time, such logics have been much discussed, and fields of applications have been sought for; the original publications left the question of application open and the writers restricted themselves to the formal construction of a calculus. The construction of a logic of probability, in which a continuous scale of truth values is introduced, has been given by the author.[4] This logic corresponds more to classical physics than to quantum mechanics. Since, in it, every proposition has a determinate probability, it has no room for a truth value of indeterminacy; a probability of $\frac{1}{2}$ is not what is meant by the category *indeterminate* of quantum mechanical statements. Probability logic is a generalization of two-valued logic for the case of a kind of truth possessing a continuous gradation. Quantum mechanics is interested in such a logic only so far as a generalization of its categories *true* and *false* is intended, which is necessary in this domain in the same sense as in classical physics; the use of the "sharp" categories *true* and *false* must be considered in both cases as an idealization applicable only in the sense of an

---

[2] E. L. Post, "Introduction to a General Theory of Elementary Propositions," *Am. Journ. of Math.*, XLIII (1921), p. 163.
[3] J. Lucasiewicz, *Comptes rendus Soc. d. Sciences Varsovie*, XXIII (1930), Cl. III, p. 51; J. Lucasiewicz and A. Tarski, *op. cit.*, p. 1. The first publication by Lucasiewicz of his ideas was made in the Polish journal *Ruch Filozoficzny*, V (Lwow, 1920), pp. 169–170.
[4] H. Reichenbach, "Wahrscheinlichkeitslogik," *Ber. d. Preuss. Akad.*, *Phys.-Math. Kl.* (Berlin, 1932).

approximation. The quantum mechanical truth-value *indeterminate*, however, represents a topologically different category. The application of a three-valued logic to quantum mechanics has been frequently envisaged; thus, Paulette Février[5] has published the outlines of such a logic. The construction which we shall present here is different, and is determined by the epistemological considerations presented in the preceding sections.

§ 31. The Rules of Two-valued Logic

Before we turn to the presentation of the system of three-valued logic, let us give a short exposition of the rules of two-valued logic. We shall use the *logistic*, or *symbolic*, form of logic, since only this form is sufficiently precise to make possible an extension to a three-valued system.

Classical, or two-valued, logic is represented, with respect to its structure, by *truth tables* which determine the truth values resulting for propositional operations as functions of those of the elementary propositions. The most important of these operations are the following:

$\bar{a}$ non-a, negation
$a \vee b$ a or b, or both, disjunction
$a \cdot b$ a and b, conjunction
$a \supset b$ a implies b
$a \equiv b$ a equivalent to b

The two-valued truth tables are set forth in tables 3A and 3B.

These tables can be read in two directions: from the elementary propositions to the propositional combination, or from the combination to the elementary propositions. For the "or", for instance, the tables tell us by the first direction: "If a is true and b is true, $a \vee b$ is true". By the second direction they tell us: "If $a \vee b$ is true, a is true and b is true, or a is true and b is false, or a is false and b is true." The more T's the column of an operation contains, the weaker is the operation, since it tells less. Thus the "or" is weaker than the "and"; it informs us to a less extent about the truth values of a and b than the "and". A weaker operation is easier to verify than a stronger one, since any of the T-cases, if observed, will verify it. For the implication of table 3B, which corresponds only to a certain extent to the implication of conversational language, Russell has introduced the name of *material implication*.

---

[5] Paulette Février, "Les relations d'incertitude de Heisenberg et la logique," *Comptes rendus de l'Acad. d. Sciences*, T. 204 (Paris, 1937), pp. 481, 958. Cf. also the report given by L. Rougier, "Les nouvelles logiques de la mécanique quantique," *Journ. of Unified Science, Erkenntnis*, Vol. 9 (1939), p. 208. In P. Février's paper the third truth value is not considered as indeterminacy, but as "reinforced falsehood" or absurdity; accordingly, her truth tables are different from ours. She makes a distinction between "propositions conjuguées" and "propositions non conjuguées" which is similar to the distinction introduced by M. Strauss. Our objections against the latter conception (p. 144) therefore apply also to this conception.

§31. RULES OF TWO-VALUED LOGIC

A logical formula is a combination of propositions which, for every truth value of the elementary propositions, has the truth value T. Such a formula is called a *tautology*. It is necessarily true, since it is true whatever be the truth values of the elementary propositions. On the other hand, a tautology is empty, or tells nothing, since it does not inform us at all about the truth values of the elementary propositions. This property does not make tautologies valueless; on the contrary, their value consists in their being necessary and empty. Such formulae can always be added to physical statements, since no empirical content is added by them; and we must add them if we want to derive conse-

TABLE 3A

a	Negation $\bar{a}$
T	F
F	T

TABLE 3B

a	b	Disjunction $a \lor b$	Conjunction $a \cdot b$	Implication $a \supset b$	Equivalence $a \equiv b$
T	T	T	T	T	T
T	F	T	F	F	F
F	T	T	F	T	F
F	F	F	F	T	T

quences from physical statements. The construction of elaborate tautologies therefore presents the physicist with a powerful instrument of derivation; the whole of mathematics must be regarded as an instrument of this kind.

Examples of simple tautologies are given by the formulae:

$a \equiv a$ rule of identity (1)
$\bar{\bar{a}} \equiv a$ rule of double negation (2)
$a \lor \bar{a}$ tertium non datur (3)
$\overline{a \cdot \bar{a}}$ rule of contradiction (4)
$\overline{a \cdot b} \equiv \bar{a} \lor \bar{b}$ } rules of De Morgan (5)
$\overline{a \lor b} \equiv \bar{a} \cdot \bar{b}$ (6)
$a \cdot (b \lor c) \equiv a.b \lor a.c$ 1st distributive rule (7)
$a \lor b.c \equiv (a \lor b).(a \lor c)$ 2nd distributive rule (8)
$(\bar{a} \supset b) \equiv \bar{b} \supset a)$ rule of contraposition (9)
$(a \equiv b) \equiv (a \supset b).(b \supset a)$ dissolution of equivalence (10)
$a \supset b \equiv \bar{a} \lor b$ dissolution of implication (11)
$(a \supset \bar{a}) \supset \bar{a}$ reductio ad absurdum (12)

150   PART III. INTERPRETATIONS

All these formulae can easily be verified by *case analysis*, i.e., by assuming for a and b successivley all truth values and proving on the basis of the truth tables that the truth value of the formula is always T. To simplify our notation we have used the following *rule of binding force* for the symbols:

$$\text{Strongest binding force} \quad ^- \;.\; \vee \;\supset\; \equiv \quad \text{weakest binding force}$$

This rule saves parentheses.

A formula which has sometimes T, sometimes F, in its column is called *synthetic*. It states an empirical truth. All physical statements, whether they be physical laws or statements about physical conditions at a given time, are synthetic. A formula which has only F's in its column is called a contradiction; it is always false.

Starting with the given tautologies, we can easily construct rules which allow us to manipulate logical formulae in a way similar to the methods of mathematics. We shall not enter into a description of this procedure here; it is explained in textbooks of symbolic logic.

§ 32. The Rules of Three-valued Logic

The method of constructing a three-valued logic is determined by the idea that the metalanguage of the language considered can be conceived as belonging to a two-valued logic. We thus consider statements of the form "A has the truth value T" as two-valued statements. The truth tables of three-valued logic then can be constructed in a way analogous to the construction of the tables of two-valued logic. The only difference is that in the vertical columns to the left of the double line we must assume all possible combinations of the three values T, I, F.

The number of definable operations is much greater in three-valued tables than in two-valued ones. The operations defined can be considered as generalizations of the operations of two-valued logic; we then, however, shall have various generalizations of each operation of two-valued logic. We thus shall obtain various forms of negations, implications, etc. We confine ourselves to the definition of the operations presented in truth tables 4A and 4B.[1] As before, three-valued propositions will be written with capital letters.

The negation is an operation which applies to one proposition; therefore only one negation exists in two-valued logic. In three-valued logic several operations applying to one proposition can be constructed. We call all of them negations because they change the truth value of a proposition. It is expedient to consider the truth values, in the order T, I, F, as running from the *highest*

---

[1] Most of these operations have been defined by Post, with the exception of the complete negation, the alternative implication, the quasi implication, and the alternative equivalence, which we introduce here for quantum mechanical purposes. Post defines some further implications which we do not use. Our standard implication is Post's implication $\supset_m^\mu$ with $m = 3$ and $\mu = 1$, i.e., for a three-valued logic and $t_\mu = t_1 =$ truth.

§32. RULES OF THREE-VALUED LOGIC

value T to the *lowest* value F. Using this terminology, we may say that the cyclical negation shifts a truth value to the next lower one, except for the case of the lowest, which is shifted to the highest value. We therefore read the expression $\sim A$ in the form *next-A*. The diametrical negation reverses T and F, but leaves I unchanged. This corresponds to the function of the arithmetical

TABLE 4A

A	Cyclical negation $\sim A$	Diametrical negation $-A$	Complete negation $\overline{A}$
T	I	F	I
I	F	I	T
F	T	T	T

TABLE 4B

A	B	Disjunction $A \vee B$	Conjunction $A . B$	Standard implication $A \supset B$	Alternative implication $A \rightarrow B$	Quasi implication $A \rightarrowtail B$	Standard equivalence $A = B$	Alternative equivalence $A \equiv B$
T	T	T	T	T	T	T	T	T
T	I	T	I	I	F	I	I	F
T	F	T	F	F	F	F	F	F
I	T	T	I	T	T	I	I	F
I	I	I	I	T	T	I	T	T
I	F	I	F	I	T	I	I	F
F	T	T	F	T	T	I	F	F
F	I	I	F	T	T	I	I	F
F	F	F	F	T	T	I	T	T

minus sign when the value I is interpreted as the number 0; and we therefore call the expression $-A$ *the negative of A*, reading it as *minus-A*. The complete negation shifts a truth value to the higher one of the other two. We read $\overline{A}$ as *non-A*. The use of this negation will become clear presently.

Disjunction and conjunction correspond to the homonymous operations of two-valued logic. The truth value of the disjunction is given by the higher one of the truth values of the elementary propositions; that of the conjunction, by the lower one.

There are many ways of constructing implications. We shall use only the three implications defined in table 4B. Our first implication is a three-three operation, i.e., it leads from three truth values of the elementary propositions to three truth values of the operation. We call it *standard implication*. Our second implication is a three-two operation, since it has only the values T and F in its column; we therefore call it *alternative implication*. Our third implication is called *quasi implication* because it does not satisfy all the requirements which are usually made for implications.

What we demand in the first place of an implication is that it makes possible the procedure of *inference*, which is represented by the rule: If A is true, and A implies B is true, then B is true. In symbols:

$$\frac{\begin{array}{c} A \\ A \supset B \end{array}}{B} \qquad (1)$$

All our three implications satisfy this rule; so will every operation which has a T in the first line and no T in the second and third line of its truth table. In the second place, we shall demand that if A is true and B is false, the implication is falsified; this requires an F in the third line—a condition also satisfied by our implications. These two conditions are equally satisfied by the "and", and we can indeed replace the implication in (1) by the conjunction. If we do not consider the "and" as an implication, this is owing to the fact that the "and" says too much. If the second line of (1) is $A.B$, the first line can be dropped, and the inference remains valid. We thus demand that the implication be so defined that without the first line in (1) the inference does not hold; this requires that there are some T's in the lines below the third line. This requirement is satisfied by the first and second implication, though not by the quasi implication. A further condition for an implication is that *a implies a* is always true. Whereas the first and second implication satisfy this condition, the quasi implication does not. The reason for considering this operation, in spite of these discrepancies, as some kind of implication will appear later (cf. § 34).

It is usually required that A implies B does not necessarily entail B implies A, i.e., that the implication is nonsymmetrical. Our three implications fulfill this requirement. The latter condition distinguishes an implication from an equivalence (and is also a further distinction from the "and"). The equivalence is an operation which states equality of truth values of A and B; it therefore must have a T in the first, the middle, and the last line. Furthermore, it is required to be symmetrical in A and B, such that with A equivalent B we also have B equivalent A. These conditions are satisfied by our two equivalences. Since these conditions leave the definition of equivalence open within a certain frame, further equivalences could be defined; we need, however, only the two given in the tables.

§32. RULES OF THREE-VALUED LOGIC 153

To simplify our notation we use the following *rule of binding force* for our symbols:

strongest binding force

complete negation	—
cyclical negation } equal force	∼
diametrical negation	–
conjunction	.
disjunction	∨
quasi implication	⇸
standard implication	⊃
alternative implication	→
standard equivalence	≡
alternative equivalence	≡

weakest binding force

If several negations of the diametrical or cyclical form precede a letter A, we convene that the one immediately preceding A has the strongest connection with A, and so on in the same order. The line of the complete negation extended over compound expressions will be used like parentheses.

Our truth values are so defined that only a statement having the truth value T can be asserted. When we wish to state that a statement has a truth value other than T, this can be done by means of the negations. Thus the assertion

$$\sim\sim A \qquad (2)$$

states that A is indeterminate. Similarly, either one of the assertions

$$\sim A \qquad -A \qquad (3)$$

states that A is false.

This use of the negations enables us to eliminate statements in the metalanguage about truth values. Thus the statement of the object language *next-next-A* takes the place of the semantical statement "A is indeterminate". Similarly, the statement of the metalanguage "A is false" is translated into one of the statements (3) of the object language, and then is pronounced, respectively, "next-A", or "minus-A". We thus can carry through the principle that *what we wish to say is said in a true statement of the object language*.

As in two-valued logic, a formula is called *tautological* if it has only T's in its column; *contradictory*, if it has only F's; and *synthetic*, if it has at least one T in its column, but also at least one other truth value. Whereas the statements of two-valued logic divide into these three classes, we have a more complicated division in three-valued logic. The three classes mentioned exist also in the three-valued logic, but between synthetic and contradictory statements we have a class of statements which are never true, but not contradictory; they have only I's and F's in their column, or even only I's, and may be called *asynthetic* statements. The class of synthetic statements subdivides into three categories. The first consists of statements which can have all three truth

154   PART III. INTERPRETATIONS

values; we shall call them *fully synthetic statements*. The second contains statements which can be only true or false; they may be called *true-false* statements, or *plain-synthetic* statements. They are synthetic in the simple sense of two-valued logic. The use of these statements in quantum mechanics will be indicated on p. 159. The third category contains statements which can be only true or indeterminate. Of the two properties of the synthetic statements of two-valued logic, the properties of being sometimes true and sometimes false, these statements possess only the first; they will therefore be called *semisynthetic* statements.

The cyclical or the diametrical negation of a contradiction is a tautology; similarly, the complete negation of an asynthetic statement is a tautology. A synthetic statement cannot be made a tautology simply by the addition of a negation.

All quantum mechanical statements are synthetic in the sense defined. They assert something about the physical world. Conversely, if a statement is to be asserted, it must have at least one value T in its column determined by the truth tables. Asserting a statement means stating that one of its T-cases holds. Contradictory and asynthetic statements are therefore *unassertable*. On the other hand, tautologies and semisynthetic statements are *indisprovable*; they cannot be false. But whereas tautologies must be true, the same does not follow for semisynthetic statements. When a semisynthetic statement is asserted, this assertion has therefore a *content*, i.e., is not *empty* as in the case of a tautology. For this reason we include semisynthetic statements in the synthetic statements; all synthetic statements, and only these, have a content.

The unique position of the truth value T confers to tautologies of the three-valued logic the same rank which is held by these formulae in two-valued logic. Such formulae are always true, since they have the value T for every combination of the truth values of the elementary propositions. As before, the proof of tautological character can be given by case analysis on the base of the truth tables; this analysis will include combinations in which the elementary propositions have the truth value I. We now shall present some of the more important tautologies of three-valued logic, following the order used in the presentation of the two-valued tautologies (1)–(12), § 31.

The *rule of identity* holds, of course:

$$A \equiv A \qquad (4)$$

The *rule of double negation* holds for the diametrical negation:

$$A \equiv --A \qquad (5)$$

For the cyclical negation we have a *rule of triple negation:*

$$A \equiv \sim\sim\sim A \qquad (6)$$

§32. RULES OF THREE-VALUED LOGIC

For the complete negation the rule of double negation holds in the form

$$\bar{\bar{A}} \equiv A \tag{7}$$

It should be noticed that from (7) the formula $A \equiv \bar{\bar{A}}$ cannot be deduced, since it is not permissible to substitute A for $\bar{A}$; and this formula, in fact, is not a tautology. We shall therefore say that the rule of double negation does not hold *directly*. A permissible substitution is given by substituting $\bar{A}$ for A; in this way we can increase the number of negation signs in (7) correspondingly on both sides. This peculiarity of the complete negation is explained by the fact that a statement above which the line of this negation is drawn is thus reduced to a semisynthetic statement; further addition of such lines will make the truth value alternate only between truth and indeterminacy.

Between the cyclical and the complete negation the following relation holds:

$$\bar{A} \equiv \sim A \vee \sim\sim A \tag{8}$$

The tertium non datur does not hold for the diametrical negation, since $A \vee -A$ is synthetic. For the cyclical negation we have a *quartum non datur*:

$$A \vee \sim A \vee \sim\sim A \tag{9}$$

The last two terms of this formula can be replaced by $\bar{A}$, according to (8); we therefore have for the complete negation a formula which we call a *pseudo tertium non datur*:

$$A \vee \bar{A} \tag{10}$$

This formula justifies the name "complete negation" and, at the same time, reveals the reason why we introduce this kind of negation; the relation (8), which makes (10) possible, may be considered as the definition of the complete negation. The name which we give to this formula is chosen in order to indicate that the formula (10) does not have the properties of the tertium non datur of two-valued logic. The reason is that the complete negation does not have the properties of an ordinary negation: It does not enable us to infer the truth value of A if we know that $\bar{A}$ is true. This is clear from (8); if we know $\bar{A}$, we know only that A is either false or indeterminate. This ambiguity finds a further expression in the fact that for the complete negation no converse operation can be defined, i.e., no operation leading from $\bar{A}$ to A. Such an operation is impossible, because its truth tables would coordinate to the value T of $\bar{A}$, sometimes the value I of A, and sometimes the value F of A.

The *rule of contradiction* holds in the following forms:

$$\overline{A.\bar{A}} \tag{11}$$

$$\overline{A.\sim A} \tag{12}$$

$$\overline{A.-A} \tag{13}$$

The *rules of De Morgan* hold only for the diametrical negation:

$$-(A.B) \equiv -A \vee -B \tag{14}$$

$$-(A \vee B) \equiv -A.-B \tag{15}$$

The two *distributive rules* hold in the same form as in two-valued logic:

$$A.(B \vee C) \equiv A.B \vee A.C \tag{16}$$

$$A \vee B.C \equiv (A \vee B).(A \vee C) \tag{17}$$

The *rule of contraposition* holds in two forms

$$-A \supset B \equiv -B \supset A \tag{18}$$

$$\overline{A} \rightarrow B \equiv \overline{B} \rightarrow A \tag{19}$$

Since for the diametrical negation the rule of double negation (5) holds, (18) can also be written in the form

$$A \supset B \equiv -B \supset -A \tag{20}$$

This follows by substituting $-A$ for A in (18). For (19), however, a similar form does not exist, since for the complete negation the rule of double negation does not hold directly.

The *dissolution of equivalence* holds in its usual form only for the standard implication in combination with the standard equivalence:

$$(A \equiv B) \equiv (A \supset B).(B \supset A) \tag{21}$$

The corresponding relation between alternative implication and alternative equivalence is of a more complicated kind:

$$(A \equiv B) \equiv (A \rightleftarrows B).(-A \rightleftarrows -B) \tag{22}$$

By the double arrow implication we mean implications in both directions. This double implication does not have the character of an equivalence, since it has values T in its column aside from the first, the middle, and the last line. By the addition of the second term these T's are eliminated, and the column of the alternative equivalence is reached. For a double standard implication, and similarly for two-valued implication, a second term of the form occurring on the right hand side of (22) is dispensable, because such a term follows from the first by means of the rule of contraposition (20). For the double alternative implication this is not the case. This relation states only that B is true if A is true, and that A is true if B is true; but it states nothing about what happens when A and B have one of the other truth values. A corresponding addition is given by the second term on the right hand side of (22).

The *dissolution of implication* holds for the alternative implication in the form

$$A \rightarrow B \equiv \sim -(\overline{A} \vee B) \tag{23}$$

§32. RULES OF THREE-VALUED LOGIC

The *reductio ad absurdum* holds in the two forms:

$$(A \supset \bar{A}) \supset \bar{A} \tag{24}$$

$$(A \rightarrow \bar{A}) \rightarrow \bar{A} \tag{25}$$

Next to the tautologies, those formulae offer a special interest which can only have two truth values. Among these the *true-false* statements, or plain-synthetic statements, are of particular importance. An example is given by the formula

$$\sim\sim(\sim A \lor \sim\sim A) \tag{26}$$

which assumes only the truth values T and F when A runs through all three truth values. The existence of such statements shows that the statements of three-valued logic contain a subclass of statements which have the two-valued character of ordinary logic. For the formulae of this subclass the tertium non datur holds with the diametrical negation. Thus, if D is a true-false formula, for instance the formula (26), the formula

$$D \lor -D \tag{27}$$

is a tautology.

The other two-valued formulae can easily be transformed into true-false formulae by the following device. An asynthetic formula A, which has the two values I and F in its truth table, can be transformed into the true-false formula $\sim A$. A semisynthetic formula A, which has the two values I and T, can be transformed into the true-false formula $\sim\sim A$.

We now turn to the formulation of complementarity. We call two statements *complementary* if they satisfy the relation

$$A \lor \sim A \rightarrow \sim\sim B \tag{28}$$

The left hand side is true when A is true and when A is false; in both these cases, therefore, the right hand side must be true. This is the case only if B is indeterminate. When A is indeterminate the left hand side is indeterminate; then we have no restriction for the right hand side, according to the definition of the alternative implication. Therefore (28) can be read: If A is true or false, B is indeterminate.

Substituting in (8) $\sim\sim A$ for A, and using (6), we derive

$$\overline{\sim\sim A} \equiv A \lor \sim A \tag{29}$$

We therefore can write (28) also in the form

$$\overline{\sim\sim A} \rightarrow \sim\sim B \tag{30}$$

Applying (19) we see that (30) is tautologically equivalent to

$$\overline{\sim\sim B} \rightarrow \sim\sim A \tag{31}$$

Substituting B for A in (29) we can transform (31) into

$$B \lor \sim B \rightarrow \sim\sim A \tag{32}$$

It follows that (32) is tautologically equivalent to (28).[2] The condition of complementarity is therefore symmetrical in A and B; *if A is complementary to B, then B is complementary to A.*

The relation of complementarity, opposing the truth value of indeterminacy to the two values of truth and falsehood, is a unique feature of three-valued logic, which has no analogue in two-valued logic. Since this relation determines a column in the truth tables of A and B, it can be considered as establishing a logical operation of complementarity between A and B, for which we could introduce a special sign. It appears, however, to be more convenient to dispense with such a special sign and to express the operation in terms of other operations, in accordance with the corresponding procedure used for certain operations of two-valued logic.

The *rule of complementarity of quantum mechanics* can now be stated as follows: If u and v are noncommutative entities, then

$$U \vee \sim U \rightarrow \sim\sim V \qquad (33)$$

Here U is an abbreviation for the statement, "The first entity has the value u"; and V for, "The second entity has the value v". Because of the symmetry of the complementarity relation, (33) can also be written

$$V \vee \sim V \rightarrow \sim\sim U \qquad (34)$$

Furthermore, the two forms (30) and (31) can be used.

With (33) and (34) we have succeeded in formulating the rule of complementarity in the object language. This rule is therefore stated as a physical law having the same form as all other physical laws. To show this let us consider as an example the law: If a physical system is closed (statement a), its energy does not change (statement $\bar{b}$). This law, which belongs to two-valued logic, is written symbolically:[3]

$$a \supset \bar{b} \qquad (35)$$

This is a statement of the same type as (33) or (34). It is therefore not necessary to read (33) in the *semantical form*: "If U is true or false, V is indeterminate." Instead, we can read (33) in the object language: "U or next-U implies next-next-V".

The law (33) of complementarity can be extended to propositional functions. Our statement U can be written in the functional form

$$Vl(e_1,t) = u \qquad (36)$$

meaning: "The value of the entity e_1 at the time t is u". The symbol "$Vl()$"

---

[2] This result could not be derived if we were to use the standard implication, instead of the alternative implication, in (28) and (32).

[3] We simplify this example. A complete notation would require the use of propositional functions.

§32. RULES OF THREE-VALUED LOGIC   159

used here is a *functor*, meaning "the value of . . . ",[4] and will be similarly applied to other entities. Then the law of complementarity can be expressed in the form:

$$(u)\ (v)\ (t)\ \{[\ Vl(e_1,t) = u]\ \vee \sim[\ Vl(e_1,t) = u] \to\ \sim\sim[\ Vl(e_2,t) = v]\} \quad (37)$$

The symbols (u), (v), (t), represent all-operators and are read, as in two-valued logic: "for all u" . . . "for all t".

The relation of complementarity is not restricted to two entities; it may hold between three or more entities. Thus the three components of the angular momentum are noncommutative, i.e., each is complementary to each of the two others (cf. p. 78). In order to express this relation for the three entities u,v,w, we add to (33) the two relations:

$$V \vee \sim V \to\ \sim\sim W \qquad W \vee \sim W \to\ \sim\sim U \quad (38)$$

Each of these relations can be reversed, as has been shown above. The three relations (33) and (38) then state that, if one of the three statements is true or false, the other two are indeterminate.

Since the alternative implication is the *major operation* in (33) and (34),[5] it follows that these formulae can only be true or false, but not indeterminate. The rule of complementarity, although it concerns all three truth values, is therefore, in itself, a true-false formula. Since the rule is maintained by quantum mechanics as true, it has the truth of a two-valued synthetic statement. We see that this interpretation of the rule of complementarity, which is implicitly contained in the usual conception of quantum mechanics, appears as a logical consequence of our three-valued interpretation.[6]

This result shows that the introduction of a third truth value does not make all statements of quantum mechanics three-valued. As pointed out above, the frame of three-valued logic is wide enough to include a class of true-false formulae. When we wish to incorporate all quantum mechanical statements

---

[4] The use of functors in three-valued logic differs from that in two-valued logic in that the existence of a determinate value designated by the functor can be asserted only when the statement (36) is true or false, whereas the indeterminacy of (36) includes the indeterminacy of a statement about the existence of a value. We shall not give here the formalization of this rule.

[5] I.e., the operation which divides these formulae into two major parts.

[6] It can be shown that the true-false character of (28) and (32) is not bound to the particular form which we gave to the arrow implication, but ensues if the following postulates are introduced: one, the relation of complementarity is symmetrical in A and B, i.e., (28) is equivalent to (32); two, if A in (28) is indeterminate, B can have any one of the three truth values; three, the implication used in (28) is verified if both implicans and implicate are true, and is falsified if the implicans is true and the implicate is false. We shall only indicate the proof here. Postulate two requires that the arrow implication have a T in the three cases where A is indeterminate; postulate three requires a T in the first value of the column of the arrow implication, and an F in the third. It turns out that with this result, seven of the nine cases of (28) are determined, and contain only T's and F's. The missing case T, F of (28) then must be equal to the case F, T, according to the first postulate, and is thus determined as F. It then can be shown that in order to furnish this result, the arrow implication must have an F in the second value from the top. With this, the last case of (28), the F,F-case, is determined as F. The arrow implication is not fully determined by the given postulates; its last three values can be arbitrarily chosen.

into three-valued logic, it will be the leading idea to put into the true-false class those statements which we call quantum mechanical *laws*. Furthermore, statements about the form of the ψ-function, and therefore about the *probabilities* of observable numerical values, will appear in this class. Only statements about these numerical values themselves have a three-valued character, determined by table 2.

§ 33. Suppression of Causal Anomalies through a Three-valued Logic

In the given formulae we have outlined the interpretation of quantum mechanics through a three-valued logic. We see that this interpretation satisfies the desires that can be justifiably expressed with respect to the logical form of a scientific theory, and at the same time remains within the limitations of knowledge drawn by the Bohr-Heisenberg interpretation. The term "meaningless statement" of the latter interpretation is replaced, in our interpretation, by the term "indeterminate statement". This has the advantage that such statements can be incorporated into the object language of physics, and that they can be combined with other statements by logical operations. Such combinations are "without danger", because they cannot be used for the derivation of undesired consequences.

Thus, the and-combination of two complementary statements can never be true. This follows in our interpretation, because the formula

$$[A \lor \sim A \rightarrow \sim\sim B] \rightarrow \overline{A.B} \tag{1}$$

is a tautology. This is not an equivalence; therefore the condition of complementarity cannot be replaced by the condition $\overline{A.B}$. But the implication (1) guarantees that two complementary statements cannot be both true. Such a combination can be false; but only if the statement about the measured entity is false. Now if a measurement of q has resulted in the value q_1, it will certainly be permissible to say that the statement "the value of q is q_2, and the value of p is p_1", is false. Similarly, it is without danger when we consider the or-combination, "the value of q is q_1 or the value of p is p_1", as true *after* a measurement of q has furnished the value q_1. With such a statement nothing is said about the value of p.

Furthermore, the reductio ad absurdum (24), § 32, or (25), § 32, cannot be used for the construction of indirect proofs. If we have proved by means of the reductio ad absurdum that $\overline{A}$ is true, we cannot infer that A is false; A can also be indeterminate. Similarly, we cannot construct a disjunctive derivation of a statement C by showing that C is true both when B is true and when B is false; we then have proved only the relation

$$B \lor -B \supset C \tag{2}$$

Since the implicans need not be true, we cannot generally infer that C must be true.

This shows clearly the difference between two-valued and three-valued logic. In two-valued logic a statement c is proved when the relation

$$b \vee \bar{b} \supset c \tag{3}$$

has been demonstrated, since here the implicans is a tautology. The analogue of (3) in three-valued logic is the relation

$$B \vee \bar{B} \supset C \tag{4}$$

which according to (8), § 32, is the same as

$$B \vee \sim B \vee \sim\sim B \supset C \tag{5}$$

If (5) is demonstrated, C is proved, since here the implicans is a tautology. But this means that in order to prove C we must prove that C is true in the three cases that B is true, false, or indeterminate. An analysis of quantum mechanics shows that such a proof cannot be given if C formulates a causal anomaly; we then can prove, not (5), but only the relation (2), or a generalization of the latter relation which we shall study presently.

For this purpose we must inquire into some properties of *disjunctions*. Let us introduce the following notation, which applies both to the two-valued and the three-valued case.

A disjunction of n terms is called *closed* if, in case $n-1$ terms are false, the n-th term must be true.

A disjunction is called *exclusive* if, in case one term is true, all the others must be false.

A disjunction is called *complete* if one of its terms must be true; or what is the same, if the disjunction is true.

For the two-valued case the first two properties are expressed by the following relations:

$$\begin{aligned} b_1 &\equiv \bar{b}_2 \cdot \bar{b}_3 \ldots \bar{b}_n \\ b_2 &\equiv \bar{b}_1 \cdot \bar{b}_3 \ldots \bar{b}_n \\ &\cdots\cdots\cdots\cdots \\ b_n &\equiv \bar{b}_1 \cdot \bar{b}_2 \ldots \bar{b}_{n-1} \end{aligned} \tag{6}$$

That a disjunction for which these relations hold, is closed, follows when we read the equivalence in (6) as an implication from right to left; that it is exclusive follows when we read the equivalence as an implication from left to right. Now it can easily be shown that if relations (6) hold, the disjunction

$$b_1 \vee b_2 \vee \ldots \vee b_n \tag{7}$$

must be true. This result can even be derived if we consider in (6) only the implications running from right to left. In other words: A two-valued disjunction which is closed is also complete, and vice versa. This is the reason that in two-valued logic the terms "closed" and "complete" need not be distinguished. Furthermore, it can be shown that one of the relations (6) can be dispensed with, since it is a consequence of the others.

For the three-valued case a closed and exclusive disjunction is given by the following relations:

$$B_1 \rightleftarrows -B_2 \cdot -B_3 \ldots -B_n$$
$$B_2 \rightleftarrows -B_1 \cdot -B_3 \ldots -B_n \qquad (8)$$
$$\ldots\ldots\ldots\ldots\ldots\ldots$$
$$B_n \rightleftarrows -B_1 \cdot -B_2 \ldots -B_{n-1}$$

As before, the closed character of the disjunction follows when we use the implications from right to left; and the exclusive character follows when we use the implications from left to right.

We now meet with an important difference from the two-valued case. From the relations (8) we cannot derive the consequence that the disjunction

$$B_1 \vee B_2 \vee \ldots \vee B_n \qquad (9)$$

must be true. This disjunction can be indeterminate. This will be the case if some of the B_i are indeterminate and the others are false. All that follows from (8) is that not all B_i can be false simultaneously; the disjunction (9) therefore cannot be false. But since it can be indeterminate, we cannot derive that the disjunction is complete. In three-valued logic we must therefore distinguish between the two properties *closed* and *complete*; a closed disjunction need not be complete. A further difference from the two-valued case is given by the fact that the conditions (8) are independent of each other, i.e., that none is dispensable. This is clear, because, if the last line in (8) is omitted, the remaining conditions would be satisfied if the $B_1 \ldots B_{n-1}$ are false and B_n is indeterminate, a solution excluded by the last line of (8).

The disjunction $B \vee -B$ is a special case of a closed and exclusive disjunction. Corresponding to (2), the proof of the relation

$$B_1 \vee B_2 \vee \ldots \vee B_n \supset C \qquad (10)$$

does not represent a proof of C if the $B_1 \ldots B_n$ constitute a closed and exclusive disjunction, since the implicans can be indeterminate. Only a complete disjunction in the implicans would lead to a proof of C.

We now shall illustrate, by an example, that the distinction of closed and complete disjunctions enables us to eliminate certain causal anomalies in quantum mechanics.

Let us consider the interference experiment of § 7 in a generalized form in which n slits $B_1 \ldots B_n$ are used. Let B_i be the statement: "The particle passes through slit B_i". After a particle has been observed on the screen we know that the disjunction $B_1 \vee B_2 \ldots \vee B_n$ is closed and exclusive; namely, we know that

§ 33. SUPPRESSION OF ANOMALIES 163

if the particle did not go through $n-1$ of the slits, it went through the n-th slit, and that if it went through one of the slits, it did not go through the others. In other words, the observation of a particle on the screen implies that relations (8) hold. But since from these relations the disjunction (9) is not derivable, we cannot maintain that this disjunction is *complete*, or *true*; all we can say is that it is *not false*. It can be indeterminate. This will be the case if no observation of the particle at one of the slits has been made.

The disjunction will also be indeterminate if an observation is made at the n-th slit with the result that the particle did not go through this slit. Such observations will, of course, disturb the interference pattern on the screen. But we are concerned so far only with the question of the truth character of the disjunction. The fact that, if observations with negative result are made in less than $n-1$ slits, the remaining slits will still produce a common interference pattern finds its expression in the logical fact that in such a case the remaining disjunction is indeterminate. Only when an observation with negative result is made at $n-1$ slits, do we know that the particle went through the n-th slit; then the disjunction will be true. On the other hand, if the particle has been observed at one slit, we know that the particle did not go through the others; the disjunction then is also true.

Although the disjunction (9) may be indeterminate, our knowledge about the relations holding between the statements $B_1 \ldots B_n$ will not be indeterminate, but true or false. This follows because relations (8), whose major operation is the alternative implication, represent true-false formulae.[1]

For the case $n = 2$, relations (8) are simplified. We then have

$$\left. \begin{array}{c} B_1 \rightleftarrows -B_2 \\ B_2 \rightleftarrows -B_1 \end{array} \right\} \qquad (11)$$

Using relations (22), § 32, and (5), § 32, we can write this in the form:

$$B_1 \equiv -B_2 \qquad (12)$$

This means that B_1 is equivalent to the negative, or the diametrical negation of B_2. This relation, which we shall call a *diametrical disjunction*, can be considered as a three-valued generalization of the exclusive "or" of two-valued logic. In the case of a diametrical disjunction we know that B_1 is true if B_2 is false, that B_1 is false if B_2 is true, and that B_1 is indeterminate if B_2 is indeterminate. This expresses precisely the physical situation of such a case. If an observation at one slit is made, the statement about the passage of the particle

---

[1] As before, the true-false character of these formulae is not introduced by us deliberately, but results from other reasons. If we were to use in (8) a double standard implication, this would represent a standard equivalence according to (21), § 32; then the only case in which the disjunction (9) is indeterminate is the case that all B_i are indeterminate. But we need also cases in which some B_i are false and the others indeterminate, for the reasons explained above.

at the other slit is no longer indeterminate, whether the result of the observation at the first slit is positive or negative. We see that this particular form of the disjunction results automatically from the general conditions (8) if we put $n = 2$.[2]

Now let us regard the bearing of this result on probability relations. Using the notation introduced in § 7, we can determine the probability that a particle leaving the source A of radiation and passing through slit B_1 or slit B_2 or . . . or slit B_n will arrive at C by the formula[3]

$$P(A.[B_1 \lor B_2 \lor \ldots \lor B_n], C) = \frac{\sum_{i=1}^{n} P(A,B_i) \cdot P(A.B_i,C)}{\sum_{i=1}^{n} P(A,B_i)} \qquad (13)$$

This formula is the mathematical expression of the principle of corpuscular superposition; it states that the statistical pattern occurring on the screen, when all slits are open simultaneously, is a superposition of the individual patterns resulting when only one slit is open. Formula (13), however, can only be applied when the two statements A and $B_1 \lor B_2 \lor \ldots \lor B_n$ are true. Now A is true, since it states that the particle came from the source A of radiation. But we saw that $B_1 \lor B_2 \lor \ldots \lor B_n$ cannot be proved as true. Therefore (13) is not applicable for the case that all slits are open. The probability holding for this case must be calculated otherwise, and is not determined by the principle of corpuscular superposition. We see that no causal anomaly is derivable.

We can interpret this elimination of the anomaly as given by the impossibility of an inference based on the implication (10) when we regard the C of this relation as meaning: The probability holding for the particle has the value (13). The inference then breaks down because our knowledge, formulated by (8), does not prove the implicans of (10) to be true.

In the example of the grating we applied the relations (8) to a case where the localization of the particle is given in terms of discrete positions. The same relations can also be applied to the case of a continuous sequence of possible positions, such as will result when we make an unprecise determination of position. We then usually say: The particle is localized inside the interval Δq, but it is unknown at which point of this interval it is. The latter addition repre-

---

[2] Dr. A. Tarski, to whom I communicated these results, has drawn my attention to the fact that it is possible also to define a similar generalization of the inclusive "or". We then replace in the column of the disjunction in table 4B the "I" of the middle row by a "T", leaving all other cases unchanged. This "almost or", as it may be called, thus means that at least one of two propositions is true or both are indeterminate. This operation can be shown to be commutative and associative, while it is not distributive or reflexive (the latter term meaning that "almost A or A" is not the same as "A"). Applied to more than two propositions, the "almost or" means: "At least one proposition is true or at least two are indeterminate". A disjunction in terms of the "almost or", therefore, represents what we called above a closed disjunction. It can be shown that if relations (8) hold, the $B_1 \ldots B_n$ will constitute a disjunction of this kind. The latter statement, of course, is not equivalent to the relations (8), but merely a consequence of the latter.
[3] Cf. the author's *Wahrscheinlichkeitslehre* (Leiden, 1935), (4), § 22.

§ 33. SUPPRESSION OF ANOMALIES   165

sents the use of an exhaustive interpretation. Within a restrictive interpretation we shall also say that the particle is localized inside the interval Δq. But we shall not use the additional statement because we cannot say that the particle is at a specific point of the interval. Rather, we shall define: The phrase "inside the interval Δq" means that when we divide this interval into n small intervals $\delta q_1 \ldots \delta q_n$ adjacent to each other, the relations (8) will hold for the statements B_i made in the form "the particle is situated in δq_i". We see that the statement "the position of the particle is measured to the exactness Δq" is thus given an interpretation in terms of three-valued logic. The statement itself is true or false, since the relations (8) are so. But since we can infer from (8) only that the disjunction is closed, although it need not be complete, the statement is not translatable into the assertion "the particle is at one and only one point of the interval Δq". The fact that the latter consequence cannot be derived makes it impossible to assert causal anomalies.

In a similar way other anomalies are ruled out. Let us consider, as a further example, the anomaly connected with potential barriers. A potential barrier is a potential field so oriented that particles running in a given direction are slowed down, as the electrons emitted from the filament of a radio tube are slowed down by a negative potential of the grid. In classical physics a particle cannot pass a potential barrier unless its kinetic energy is at least equal to the additional potential energy H_o which the particle would have acquired in running up to the maximum of the barrier. In quantum mechanics it can be shown that particles which, if measured inside the barrier, possess a kinetic energy smaller than H_o can later be found with a certain probability outside the barrier. This result is not only a consequence of the mathematics of quantum mechanics, derivable even in so simple a case as a linear oscillator, but its validity is, according to Gamow, proved by the rules of radioactive disintegration. It is important to realize that the paradox cannot be eliminated by a suitable assumption about a disturbance through the measurement. Let us consider a swarm of particles having the same energy H, with $H < H_o$. That each of these particles has this energy can be shown by an energy measurement applied to each, or by taking fair samples out of a swarm originating in sufficiently homogeneous conditions. According to the considerations given on page 140, we must consider the measured value H as the value of the energy *after* the measurement. After passing through the zone of measurement the particles enter the field of the potential barrier. Beyond the barrier, even at a great distance from it, measurements of position are made which localize particles at that place. Because of the distance, the latter measurements cannot have introduced an additional energy into the particle before it reached the barrier; this means we cannot assume that the measurement of position pushes the particle across the barrier, since such an assumption would itself represent a causal anomaly, an action at a distance. We must rather say that the paradox constitutes an intrinsic difficulty of the corpuscle interpretation; it is one of the cases in which the corpuscle interpretation cannot be carried through without anomalies. The anomaly in this case represents a violation of the principle of the conservation of energy, since we cannot say that in passing the barrier the particle possesses a negative kinetic energy. In view of the fact that the kinetic energy is determined by the square of the velocity, such an assumption would lead to an imaginary velocity of the particle, a consequence incompatible with the spatio-temporal nature of particles.

If, however, we use the restrictive interpretation by a three-valued logic, this causal anomaly cannot be stated. The principle requiring that the sum of kinetic and potential energy be constant connects simultaneous values of momentum and position. If one of the two is measured, a statement about the other entity must be indeterminate, and therefore a statement about the sum of the two values will also be indeterminate. It follows that the principle of conservation of energy is eliminated, by the restrictive interpretation, from the domain of true statements, without being transformed into a false statement; it is an indeterminate statement.

What makes the paradox appear strange is this: It seems that we need not make a measurement of velocity in order to know that in passing the barrier the particle violates the principle of energy. If only we know that the velocity at this point is any real number, zero included, it follows that the principle of energy is violated. The mistake in this inference originates from the assumption, discarded by the restrictive interpretation, that an unmeasured velocity must at least have one determinate real number as its value. It is true that we know the velocity cannot be an imaginary number; but from this we can only infer that the statement, "the velocity has one real number as its value", is not false. The statement, however, will be indeterminate if the velocity is not measured. This is clear when we consider the statement as given by the closed and exclusive disjunction, "the value of the velocity is v_1 or v_2 or ...",[4] which is indeterminate for the reasons explained in the preceding example.

We see that a three-valued logic is the adequate form of a system of quantum mechanics in which no causal anomalies can be derived.

§ 34. Indeterminacy in the Object Language

We said above that the observational language of quantum mechanics is two-valued. Although this is valid on the whole, it needs some correction. We shall see this when we consider a question concerning a test of predictions based on probabilities, such as raised in § 30. For such questions the relation of complementarity introduces an indeterminacy even into the observational language.

Let us consider the two statements of observational language: "If a measure-

---

[4] It would be more correct to speak here of an existential statement instead of a disjunction of an infinite number of terms. It is clear, however, that the considerations given can be equally carried through for existential statements.

§34. INDETERMINACY IN THE OBJECT LANGUAGE   167

ment m_q is made, the indicator will show the value q_1", and "if a measurement m_p is made, the indicator will show the value p_1". We know that it is not possible to verify both these statements for simultaneous values. The case is different from the example given in § 30 concerning a throw of the die by Peter or by John; we said that in the latter case the statement about a throw of Peter's can be verified in principle even if Peter does not throw the die, by means of physical observations of another kind. For the combination of the two observational statements about measurements, however, a verification is not even possible in principle. We must therefore admit that we have in the observational language complementary statements.

Now the complementary statements of the observational language are *not* given by the two statements, "the indicator will show the value q_1", and "the indicator will show the value p_1". These statements are both verifiable, since the indicator will, or will not, show the said value even if the measurement is not made. It is rather the implications m_q *implies* q_1 and m_p *implies* p_1 which are complementary. We therefore have here in the observational language an implication which is three-valued and can have the truth-value indeterminate.

What is the nature of this implication? It is certainly not the material implication of the two-valued truth table (table 3B, p. 149), since this implication is true when the implicans is false. Thus, the statement

$$m_p \supset p_1 \tag{1}$$

conceived as written in terms of the material implication, will be true if a measurement m_q is made, since, then, the statement m_p is false. This difficulty cannot be eliminated by the attempt to interpret the implication (1) as a *tautological implication* or as a *nomological implication*, i.e., the implication of physical laws.[1] Although such an interpretation has turned out satisfactory for other cases in which the material implication appears unreasonable, it cannot be used with respect to (1), since the implication of this formula carries no necessity with it.

Now if we try to interpret the implication of (1) by the standard or the alternative implication of three-valued logic, the same difficulties as in the case of the two-valued material implication obtain. Since both m_p and p_1 are two-valued statements, we can use in the three-valued truth table (table 4B, p. 151) only those lines which do not contain the value I in the first two columns; but for these lines both the first and second implication coincide with the material implication of the two-valued truth table (table 3B). There remains therefore only the quasi implication, and we find that we must write instead of (1) the relation

$$m_p \twoheadrightarrow p_1 \tag{2}$$

This implication has the desired properties, since by canceling all lines con-

[1] A complete definition of nomological implication will be given in a later publication by the author.

taining an I in the first two columns of the table, we obtain the implication noted in table 5. Therefore, (2) corresponds to what we want to say, since we consider the statement m_p *implies* p_1 as verified or falsified only if m_p is true, whereas we consider it as indeterminate if m_p is false.

This shows that the observational language of quantum mechanics is not two-valued throughout. Although the elementary statements are two-valued, this language contains combinations of such statements which are three-valued, namely, the combinations established by the quasi implication. The truth table 3B of two-valued logic (p. 149) must therefore be complemented by the three-valued truth table (table 5) of quasi implication.[2]

TABLE 5

a	b	Quasi implication $a \looparrowright b$
T	T	T
T	F	F
F	T	I
F	F	I

We see that the three-valued logical structure of quantum mechanics penetrates to a small extent even into the observational language. Although the observational language of quantum mechanics is statistically complete, it is incomplete with respect to strict determinations. It contains a three-valued implication. If there were no relation of indeterminacy in the microcosm, this three-valued implication could be eliminated; the implication in "m_q implies q_1" then could be interpreted as a nomological implication which, in principle, could be verified or falsified. In observational relations of the kind considered, however, the uncertainty of the microcosm penetrates into the macrocosm. The same holds for all other arrangements in which an atomic occurrence releases macrocosmic processes. Such arrangements need not be measurements; they may consist as well in the lighting of lamps, or the throwing of bombs. The fact that no strict predictions can be made in microcosmic dimensions thus leads to a revision of the logical structure of the macrocosm.

[2] The quasi implication of the latter table is identical with an operation which has been introduced by the author, by the use of the same symbol, in the frame of probability logic (cf. *Wahrscheinlichkeitslehre* (Leiden, 1935), p. 381, table IIc). It can be considered as the limiting case of a probability implication resulting when only the probabilities 1 and 0 can be assumed. It can also be considered as the individual operation coordinated to probability implication as a general operation; the probability then is determined by counting only the T-cases and F-cases of the quasi implication, the I-cases being omitted. In this sense the quasi implication has been used, under the name of comma-operation, or operation of selection, in my paper "Ueber die semantische und die Objectauffassung von Wahrscheinlichkeitsausdrücken," *Journ. of Unified Science, Erkenntnis*, VIII (1939), pp. 61–62.

§ 35. The Limitation of Measurability

Our considerations concerning exhaustive and restrictive interpretations lead to a revision of the formulation given to the principle of indeterminacy. We stated this principle as a limitation holding for the measurement of simultaneous values of parameters; this corresponds to the form in which the principle has been stated by Heisenberg. We now must discuss the question whether the principle, in this formulation, is based on an exhaustive or a restrictive interpretation.

If we start from a general situation s, and consider the two probability distributions $d(q)$ and $d(p)$ belonging to s, these distributions refer to the results of measurements made in systems of the type s. Therefore, if we apply definition 4, § 29, i.e., if we regard the measured values as holding only after the measurement, the obtained values do not mean values existing in the situation s, and thus the two distributions are not referred to numerical values belonging to the same situation. The values q and p, to which these distributions refer, rather pertain, respectively, to the two different situations m_q and m_p. We therefore cannot say that the inverse correlation holding for the distributions $d(q)$ and $d(p)$ states a limitation of simultaneous values; instead, the relation of uncertainty then must be formulated as a limitation holding for values obtainable in two different situations. It follows that the usual interpretation of Heisenberg's principle as a limitation holding for the measurement of simultaneous values presupposes, not definition 4, § 29, but definition 1, § 25, since only the use of this definition enables us to interpret the results of measurements m_q and m_p as holding before the operation of measurement and thus as holding for the situation s. Therefore, if Heisenberg's inequality (2), § 3, is to be regarded as a cross-section law limiting the measurability of simultaneous values of parameters, this interpretation is based on the exhaustive interpretation expressed in definition 1, § 25.

Now we saw in § 25 that if the latter definition is assumed, we can speak of an exact ascertainment of simultaneous values when we consider a situation between two measurements and add the restriction that the situation to which the obtained combination of values belongs no longer exists at the moment when the values are known. For such values, therefore, Heisenberg's principle does not hold. It follows that the principle of uncertainty must be formulated with a qualification: The principle states a limitation holding for the measurability of simultaneous values *existing at the time when we have knowledge of them*. There is no limitation for the measurement of *past* values; only *present* values cannot be measured exactly, but are bound to Heisenberg's inequality (2), § 3. The limitation so qualified is, of course, sufficient to limit the predictability of future states; for the knowledge of past values cannot be used for predictions.

These considerations show that the conception of Heisenberg's principle as a limitation of measurability must be incorporated into an exhaustive interpretation. Within a restrictive interpretation we cannot speak of a limitation of exactness, since then the standard deviations Δq and Δp, used in the inequality (2), § 3, are not referred to the same situation. Within such an interpretation we must say that, if we regard a situation for which q is known to a smaller or greater degree of exactness Δq, p is completely unknown for this situation and cannot even be said to be at least within the interval Δp coordinated to Δq by the Heisenberg inequality. The corresponding statement holds for the reversed case. By *completely unknown* we mean here, either the category *indeterminate* of our three-valued interpretation, or the category *meaningless* of the Bohr-Heisenberg interpretation. We see that for a restrictive interpretation Heisenberg's principle, in its usual meaning, must be abandoned.

§ 36. Correlated Systems

In an interesting paper A. Einstein, B. Podolsky, and N. Rosen[1] have attempted to show that if some apparently plausible assumptions concerning the meaning of the term "physical reality" are made, complementary entities must have reality at the same time, although not both their values can be known. This paper has raised a stimulating controversy about the philosophical interpretation of quantum mechanics. N. Bohr[2] has presented his views on the subject on the basis of his principle of complementarity with the intention of showing that the argument of the paper is not conclusive. E. Schrödinger[3] has been induced to present his own rather skeptical views on the interpretation of the formalism of quantum mechanics. Other authors have made further contributions to the discussion.

In the present section we intend to show that the issues of this controversy can be clearly stated without any metaphysical assumptions, when we use the conceptions developed in this inquiry; it then is easy to give an answer to the questions raised.

In their paper Einstein, Podolsky, and Rosen construct a special kind of physical systems which may be called *correlated systems*. These are given by systems which for some time have been in physical interaction, but are separated later. They then remain correlated in such a way that the measurement of an entity u in one system determines the value of an entity v in the other system, although the latter system is not physically influenced by the act of measurement. The authors believe that this fact proves an independent reality

---

[1] A. Einstein, B. Podolsky, N. Rosen, "Can Quantum Mechanical Description of Physical Reality Be Considered Complete?" *Phys. Rev.* 47 (1935), p. 777.
[2] N. Bohr, "Can Quantum Mechanical Description of Physical Reality Be Considered Complete?" *Phys. Rev.* 48 (1935), p. 696.
[3] E. Schrödinger, "Die gegenwärtige Situation in der Quantenmechanik," *Naturwissenschaften* 23 (1935), pp. 807, 823, 844. We shall use the term "correlated systems" as a translation of Schrödinger's "verschränkte Systeme".

§ 36. CORRELATED SYSTEMS

of the entity u considered. This result appears even more plausible by a proof, given in the paper, stating that an equal correlation holds for entities other than u in the same systems, including noncommutative entities.

Translating these statements into our terminology, we can interpret the thesis of the paper as meaning that by means of correlated systems a proof can be given for the necessity of definition 1, § 25, which states that the measured value holds before and after the measurement. If we refuse to consider the measured value as holding before the measurement, such as does the Bohr-Heisenberg interpretation with definition 4, § 29, we are led to causal anomalies, since, then, a measurement in one system would physically produce the value of an entity in another system which is physically not affected by the measuring operations. This is what is claimed to be proved in the paper.

In order to analyze this argument, let us first consider the mathematical form in which it is presented. Let us assume two particles which for some time enter into an interaction; their ψ-function will then be a function $\psi(q_1 \ldots q_6)$ of six coordinates, which include the 3-position coordinates of each particle. When, after the interaction, the particles separate, the ψ-function will be given by a product of ψ-functions of the individual particles (cf. (12), § 27). Let us expand the individual ψ-functions in eigen-functions φ_i of the same entity u; we then have

$$\psi(q_1 \ldots q_6) = \sum_i \sum_k \sigma_{ik} \varphi_i(q_1, q_2, q_3) \varphi_k(q_4, q_5, q_6) \quad (1)$$

Here the σ_{ik} determine the probability $d(u_i, u_k)$ that the value u_i is measured in the first system *and* the value u_k is measured in the second system:

$$d(u_i, u_k) = |\sigma_{ik}|^2 \quad (2)$$

We have, of course,

$$\sum_i \sum_k |\sigma_{ik}|^2 = 1 \quad (3)$$

Now let us assume that a measurement of u is made in the first system, and furnishes the value u_1. The subscript 1 is not meant here to denote the first or "lowest" eigen-value, but the value obtained in the measurement. We then shall have a new ψ-function such that

$$\sum_k |\sigma_{ik}|^2 = \begin{cases} 1 & \text{for } i = 1 \\ 0 & \text{for } i \neq 1 \end{cases} \quad (4)$$

Since the second system is not involved in the measurement, its part in (1) remains unchanged; therefore the new ψ-function results from (1) simply by canceling all terms possessing a σ_{ik} with $i \neq 1$. Thus, the new ψ-function will have the form

$$\psi(q_1 \ldots q_6) = \sum_k \sigma_{1k}\varphi_1(q_1, q_2, q_3)\varphi_k(q_4, q_5, q_6)$$
$$= \varphi_1(q_1, q_2, q_3) \cdot \sum_k \sigma_{1k}\varphi_k(q_4, q_5, q_6) \quad (5)$$

with

$$\sum_k |\sigma_{1k}|^2 = \sum_k d(u_1, u_k) = d(u_1) = 1 \quad (6)$$

Let us put

$$\tau_{12}\chi_2(q_4, q_5, q_6) = \sum_k \sigma_{1k}\varphi_k(q_4, q_5, q_6) \quad (7)$$

Then (5) assumes the form

$$\psi(q_1 \ldots q_6) = \tau_{12}\varphi_1(q_1, q_2, q_3)\chi_2(q_4, q_5, q_6) \quad (8)$$

This introduction of a new ψ-function is sometimes called the *reduction of the wave packet*. Now the physical conditions of the systems, and of the measurement of u in the first system, can be so chosen that $\chi_2(q_4, q_5, q_6)$ represents an eigen-function of an entity v. Then (8) represents a situation which is determined both in u and v. This means that the situation depicted by (8) corresponds to a situation which would result from measurements of both u and v, although only a measurement of u has been made. We therefore know: If we were to measure v in the second system, we would obtain the value v_2.

We can simplify the consideration by choosing the entity v identical with the entity u. It can be proved that this is physically possible. This means that it is possible to construct physical conditions such that, after a measurement of u in the first system, a ψ-function results for which all coefficients σ_{ik} of (1) vanish except for the value σ_{12}. We then have

$$\psi(q_1 \ldots q_6) = \sigma_{12}\varphi_1(q_1, q_2, q_3)\varphi_2(q_4, q_5, q_6) \qquad |\sigma_{12}|^2 = 1 \quad (9)$$

Here the measurement of u in the first system, resulting in u_1, has made the second system definite in u, for the value u_2; i.e., if we were to measure u in the second system we would obtain the value u_2.

We now see the way the conclusion of Einstein, Podolsky, and Rosen is introduced: We must assume that the value u_2 exists in the second system *before* a measurement of u is made in this system; otherwise we are led to the consequence that a measurement of u in the first system *produces*, not only the value u_1 in the first system, but also the value u_2 in the second system. This would represent a causal anomaly, an action at a distance, since the measurement in the first system does not physically affect the second system.

With this inference the main thesis of the paper is derived. It then goes on to show that similar results can be obtained for an entity w which is comple-

§36. CORRELATED SYSTEMS

mentary to u. Let ω be the eigen-functions of w; then it is possible to make, instead of a measurement of u, a measurement of w in the first system which results in the production of an eigen-function

$$\psi(q_1 \ldots q_6) = \rho_{12} \cdot \omega_1(q_1, q_2, q_3) \cdot \omega_2(q_4, q_5, q_6) \tag{10}$$

We therefore have a free choice, either to measure u in the first system and thus to make the second system definite in u, or to measure w in the first system and to make the second system definite in w. This is considered as further evidence for the assumption that the value of an entity must exist before the measurement.

In his reply to the Einstein-Podolsky-Rosen paper, Nils Bohr sets forth the opinion that an assumption of this kind is illegitimate. Within this exposition he gives a physical interpretation of the formalism developed for correlated systems, i.e., of the above formulae (1)–(10). Let us consider this illustration before we turn to a logical analysis of the problem.

Bohr assumes that the systems considered consist of two particles, each of which passes through a slit in the same diaphragm. If we include measurements concerning the diaphragm, he continues, it is possible to determine momentum or position of the second particle by measurements of the first particle *after* the particles have passed through the slits. For the determination of the momentum of the second particle, we would have to measure:

1) the momentum of each particle *before* the particles hit the diaphragm
2) the momentum of the diaphragm *before* the particles hit
3) the momentum of the diaphragm *after* the particles hit
4) the momentum of the first particle *after* the particle hit the diaphragm

The momentum of the second particle is then determined by subtracting the change in the momentum of the first particle from the change in the momentum of the diaphragm, and adding this result to the initial momentum of the second particle.

In order to determine the position of the second particle, we would measure:

1) the distance between the slits in the diaphragm
2) the position of the first particle immediately after passing the slit

From the second result we here would infer the position of the diaphragm, which is determined because the position of the particle tells us the position of the slit through which it went (we consider here the position of the plane of the diaphragm as known, and allow only a shifting of the diaphragm within its plane, caused by the impact of the particles). Since the distance between the slits is known, the position of the second slit and with this the position of the second particle in passing, or immediately after passing the slit, is determined.

174   PART III. INTERPRETATIONS

It is interesting to see that Nils Bohr in these derivations uses the very definition which Einstein, Podolsky, and Rosen want to prove as necessary, namely, our definition 1, § 25. This definition, although not assumed for the measurements 1 and 2 of our first list, is assumed for 3 and 4, and likewise for the measurement 2 of our second list. Otherwise, for instance, the difference between the measurements 2 and 3 of the momentum of the diaphragm could not be interpreted as equal to the amount of momentum which the diaphragm has received through the impacts of both particles. If the measurement 3 changes the momentum of the diaphragm, the latter inference could not be made. Bohr does not mention his use of a definition which considers the measured value as holding before the measurement.[4] Fortunately, the use of this definition does not make Bohr's argument contradictory, as can be seen when we incorporate his answer into an analysis given in terms of our conception.

Using this conception, we shall answer the criticism of Einstein, Podolsky, and Rosen in a way different from Bohr's consideration. We shall not maintain that the use of definition 1, § 25, is *impermissible*; we shall say, instead, that this definition is *not necessary*. It may be used; thereby the corpuscle interpretation is introduced, and in the case of correlated systems of the kind considered it is this interpretation which is free from causal anomalies. Thus, even Bohr uses this exhaustive interpretation which makes his inferences plausible; he follows here the well-established habit of the physicist of switching over to the interpretation which is free from anomalies. What can be derived in such an interpretation must hold for all interpretations; this is the principle at the base of Bohr's inferences.

It would be wrong, however, to infer that because definition 1, § 25, furnishes here an interpretation free from anomalies, this definition *must* be chosen. We should be glad if we could identify this view of the problem, resulting from our inquiries, with Bohr's opinions. The latter do not seem to us to be stated sufficiently clearly to admit of an unambiguous interpretation; in particular, we should prefer to disregard Bohr's ideas about the arbitrariness of the separation into subject and object, which do not appear to us relevant for the logical problems of quantum mechanics. Let us therefore continue the analysis by the use of our own notation, applying the three-valued logic developed in § 32.

What is proved by the existence of correlated systems is that it is not permissible to say that the value of an entity before the measurement is different from the value resulting in the measurement. Such a statement would lead to

---

[4] Once this definition is chosen, the correlated systems can even be used for a measurement of simultaneous values of noncommutative entities. We then measure u in the first system, and w in the second; then the obtained values u_i and w_k represent simultaneous values in the second system. But the possibility of measuring such simultaneous values has been pointed out already in § 25 as being a consequence of definition 1, § 25. If only definition 4, § 29, is used, the two measurements on correlated systems would not furnish simultaneous values.

§ 36. CORRELATED SYSTEMS 175

causal anomalies, since it would involve the consequence that a measurement on one system produces the value of an entity in a system which is in no physical interaction with the measuring operation. In the paper of Einstein, Podolsky, and Rosen, the inference is now made that we must say that the entity before the measurement has the same value as that found in the measurement. It is this inference which is invalid.

The inference under consideration would hold only in a two-valued logic; in a three-valued logic, however, it cannot be made. Let us denote by A the statement "the value of the entity before the measurement is different from the value resulting in the measurement"; then what the existence of correlated systems proves is that, if causal anomalies are to be avoided, the statement $\bar{A}$ must hold. This statement $\bar{A}$, which states that A is not true, does not mean, however, that A is false; A can also be indeterminate. The statement, "the value of the entity before the measurement is equal to the value resulting in the measurement", is to be denoted by $-A$, i.e., by means of the diametrical negation, since this statement is true when A is false. If we could infer from the existence of correlated systems the statement $-A$, the restrictive interpretation would indeed be shown to be contradictory. But this is not the case; all that can be inferred is $\bar{A}$, and such a statement is compatible with the restrictive interpretation, since it leaves open the possibility that A is indeterminate.

We see that the paper of Einstein, Podolsky, and Rosen leads to an important clarification of the nature of restrictive interpretations. It is not permissible to understand the restrictive definition 4, § 29, as meaning that the value of the entity before the measurement is *different* from the results of the measurement; such a statement leads to the same difficulties as a statement about this value being *equal* to the result of the measurement. Every statement determining the value of the entity before the measurement will lead to causal anomalies, though these anomalies will appear in different places according as the determination of the value before the measurement is given. The anomalies appearing if equality of the values is stated are described in the interference experiment of § 7;[5] the anomalies resulting if difference of the values is stated are given in the case of correlated systems.

The given considerations constitute an instructive example for the nature of interpretations. They show the working of an exhaustive interpretation, and make it clear that restrictive interpretations are introduced for the purpose of avoiding causal anomalies; they prove, on the other hand, that the

---

[5] This is to be understood as follows. When we put a Geiger counter at the place of each of the two slits B_1 and B_2 we shall always locate the particle either in one or in the other of these two counters. (This measurement, of course, disturbs the interference pattern on the screen.) The assumption that the particle was at that place before it hits the counter leads to the consequence that the particle would have been there also when no measurement was made. This result implies that, when no observation at the slits is made, the particle will go either through one or the other slit. We showed in § 7 that this assumption leads to causal anomalies.

restrictive interpretation is consistent if all statements involved are dealt with by the rules of a three-valued logic.

Within the frame of the restrictive interpretation it is even possible to express the condition of correlation holding between the two systems after their interaction has been terminated. Using the functor $Vl()$ introduced on page 159 and indicating the system I or II within the parentheses of this symbol we can write:

$$(u)\{[Vl(e_1,I) = u] \equiv [Vl(e_1,II) = f(u)]\} \qquad (11)$$

where the function f is known. Similarly we can write for a noncommutative entity v:

$$(v)\{[Vl(e_2,I) = v] \equiv [Vl(e_2,II) = g(v)]\} \qquad (12)$$

where the function g is known. Both relations hold so long as no measurement is made in one of the systems. After a measurement has been made in one of them only that relation continues to hold which concerns the measured entity. For instance, if u has been measured in system I, only (11) continues to hold.

It would be a mistake to infer from (11) or (12) that there exists a determinate value of u, or v, in one of the systems so long as no measurement is made. This would mean that the expressions in the brackets must be true or false; but the equivalence will hold also if these expressions are indeterminate. In (11) and (12) we thus have a means of expressing the correlations of the systems without stating that determinate values of the respective entities exist.[6] This represents an advantage of the interpretation by a three-valued logic over the Bohr-Heisenberg interpretation. For the latter interpretation the statements (11)–(12) would be meaningless. Only the three-valued logic gives us the means to state the correlation of the systems as a condition holding even before a measurement is made, and thus to eliminate all causal anomalies. We need not say that the measurement of u in system I produces the value of u in the distant system II. The predictability of the value u_2 in the system II, after u has been measured in the system I, appears as a consequence of condition (11), which in its turn is a consequence of the common history of the systems.

§ 37. Conclusion

Combining the general considerations of Part I with the mathematical and logical analysis of Parts II and III, we can summarize the results of our inquiry as follows. The relation of indeterminacy is a fundamental physical law; it holds for all possible physical situations and therefore involves a disturbance of the object by the measurement. Since the relation of indeterminacy makes it impossible to verify statements about the simultaneous values of complementary entities, such statements can be introduced only by means of definitions. The physical world therefore subdivides into the world of phenomena,

[6] Cf. fn. 4, p. 159.

§ 37. CONCLUSION

which are inferable from observations in a rather simple way and therefore can be called observable in a wider sense; and the world of interphenomena, which can be introduced only by an interpolation based on definitions. It turns out that such a supplementation of the world of phenomena cannot be constructed free from anomalies. This result is not a consequence of the principle of indeterminacy; it must be considered as a second fundamental law of the physical world, which we call the principle of anomaly. Both these principles are derivable from the basic principles of quantum mechanics.

Instead of speaking of the structure of the physical world, we may consider the structure of the languages in which this world can be described; such analysis expresses the structure of the world indirectly, but in a more precise way. We then distinguish between observational language and quantum mechanical language. The first shows practically no anomalies with the exception of unverifiable implications occurring in some places. Quantum mechanical language can be formulated in different versions; we use in particular three versions: the corpuscle language, the wave language, and a neutral language. All three of these languages concern phenomena and interphenomena, but each of them shows a characteristic deficiency. Both the corpuscle language and the wave language show a deficiency so far as they include statements of causal anomalies, which occur in places not corresponding to each other and therefore can be transformed away, for every physical problem, by choosing the suitable one of the two languages. The neutral language is neither a corpuscle language nor a wave language, and thus does not include statements expressing causal anomalies. The deficiency reappears here, however, through the fact that the neutral language is three-valued; statements about interphenomena obtain the truth-value *indeterminate*.

The stated deficiencies are not due to an inappropriate choice of these languages; on the contrary, these three languages represent optima with respect to the class of all possible languages of quantum mechanics. The deficiencies must rather be regarded as the linguistic expression of the structure of the atomic world, which thus is recognized as intrinsically different from the macro-world, and likewise from the atomic world which classical physics had imagined.

INDEX

(The numbers refer to pages)

action by contact, 29
active interpretation of a transformation, 56
adjoint matrix, 58
anomaly, causal, 26, 28, 43, 160
anomaly, principle of, 33, 44, 117, 129
assumption Γ, 127
asynthetic statements, 153

Bargmann, V., 92
basic functions of an expansion, 45
basic principles of quantum mechanical method, 109, 110
binding force, rule of, in two-valued logic, 150; in three-valued logic, 153
Bohr, N., 22, 40, 170, 173
Bohr-Heisenberg interpretation, 22, 139
Boltzmann, L., 1
Born, M., 8, 22, 80, 82
Born's statistical interpretation of the waves, 36
Broglie, L. de, 6, 8, 21, 32, 66, 80

canonically conjugated parameters, 77, 100
Carnap, R., 136, 141
case analysis, 150
causal anomaly, 26, 28, 43, 160; eliminability of, 34
causal chain, 117, 122
chain structure, 123
closed disjunction, 161
commutation rule, 77
commutative operators, 76
commutator, 76
complementarity, principle of, 22
complementary parameters, 77
complementary statements, 157
complete disjunction, 161
complete set of basic functions, 45
complex functions, 45
complex vectors, 54
concentrated distribution, 96
configuration space, 65
conservation of energy, 89, 165
context, physical, 81
context of discovery or of justification, 67
continuous case of an expansion, 49

contradiction, 150, 153, 155
contraposition, rule of, 156
converse of a transformation, 58
corpuscle interpretation, 21, 25, 118
corpuscle language, 146
correlated systems, 170
correlation, inverse, 10
cross-section law, 4, 16

data, observable, 5
Davisson, C. J., 21
definite, 97
definition 1, 118; definition 2, 120; definition 3, 130; definition 4, 140; definition 5, 141; definition 6, 144
De Morgan, rules of, 156
density function, 51
derivative relations of classical physics, 111; of classical-statistical physics, 112; of quantum mechanics, 115
descriptions, equivalent, 19
descriptive simplicity, 20
determinism, 1
diametrical disjunction, 163
Dirac, P. A. M., 80
Dirac function, 51, 79
discovery, context of, 67
disjunction, closed, 161; complete, 161; diametrical, 163; exclusive, 161
dispersion, law of, 69
dissolution of equivalence, 156
dissolution of implication, 156
distribution, concentrated, 96; probability, 5, 91, 111
distributive rule, 156
disturbance by the means of observation, 15, 99, 103, 104
dualism of waves and corpuscles, 68
duality of interpretations, 33, 71
dynamic parameters, 11

eigen-functions, 48, 73
eigen-functions (practical), 97
eigen-values, 73
eigen-values, rule of, 82
Einstein, A., 8, 14, 15, 20, 21, 26, 67, 170
eliminability of causal anomalies, 34

INDEX

elimination, rule of, 28, 101
enumeration, induction by, 105
equivalence, dissolution of, 156
equivalent descriptions, 19
excluded middle, principle of, 145
exclusive disjunction, 161
exhaustive interpretations, 33, 139
Exner, F., 1
expansion, Fourier, 7, 48
expansion of a function, 45
expectation formula, 85

Feenberg, E., 92, 131
Fourier expansion, 7, 48
Fourier functions, 51
fully synthetic statements, 154
function space, 57
functions, basic, 45; complex, 45
functor, 159
fundamental tone, 7

Gauss function, 6
Germer, L. H., 21

Heisenberg, W., 3, 12, 40, 66, 80
Hermitean operators, 74
heterogeneous case (of an expansion), 49
Hilbert space, 55
holistic transformation, 65
homogeneous case (of an expansion), 49
Huygens, C., 6, 21
Huygens, principle of, 26

identity, rule of, 154
implication, dissolution of, 156
implication, material, 2, 148
implication, material, in three-valued logic, 152, 159
implication, nomological, 167; probability, 2
indeterminacy, principle of, 3, 9, 14, 35, 44, 98; (truth value), 42, 145, 146
induction by enumeration, 105
inductive simplicity, 20
inference, rule of, 152; statistical, 105
inner product, 52
integrable, quadratically, 46
integral transformation, 65
interference experiment, 24, 162
interference of probabilities, 105
interphenomena, 21

interpretation, active or passive, of a transformation, 56; Bohr-Heisenberg, 22, 139; corpuscle, 21, 25, 118; duality of, 33, 71; exhaustive, 33, 139; restrictive, 33, 40, 139; wave, 26, 129
interpretation by a restricted meaning, 141
inverse correlation, 10, 11

Jordan, P., 80
justification, context of, 67

Kemble, E. C., 92
kinematic parameters, 11
Klein, F., 56
Kramers, H. A., 78, 91

Landé, A., 23, 25
Laue, M. von, 21
law, cross-section, 4, 16
laws, causal or statistical, 1
linear operators, 74
Liouville, theorem of, 113
logic, multivalued, 147; three-valued, 42, 147, 150; two-valued, 42, 148
Lucasiewicz, J., 147

macrocosmic analogy, 39
major operation, 159
material implication, 148
matrix, 48; adjoint, 58; operator, 78
matrix mechanics, 79
matrix multiplication, 59
maximal measurement, 100
meaning, probability, 121; restricted, 42, 141; truth, 121; verifiability theory of, 29
measurability, limitation of, 4, 13, 169
measurement, general definition of, 95; quantum mechanical definition of, 97; disturbance by the, 99, 103, 104; maximal, 100
metalanguage, 141
minimum system, 33
mixture (statistical assemblages), 106
momentum, 5
monochromatic waves, 7
Morris, C. W., 141
multivalued logic, 147

needle radiation, 21
negation, rule of double, 154, 155; rule of triple, 154

Neumann, J. von, 14
neutral language, 146
Newton, I., 21
nomological implication, 167
noncommutative operators, 76
nonconnectable statements, 144
normal system, 19, 20; in the wider sense, 24
normalized functions, 46, 81

object language, 141
observable data, 5
observation, disturbance by the, 15, 104
observational language, 136
operator, 56, 72
operator matrix, 78
operators, commutative, 76; linear and Hermitean, 74
orthogonal coordinates on the sphere, 34
orthogonal straight-line coordinates, 20
orthogonality of basic functions, 45; of transformations, 52, 53
overtones, 7

passive interpretation, 56
Pauli, W., 23, 100
phenomena, 21
physical context, 81
physical situation, 81, 108
pilot waves, 32
plain-synthetic statements, 154
Planck, M., 8, 21, 67
Planck relation, 12
Podolsky, B., 170
point transformation, 65
Post, E. L., 147, 150
potential barriers, 165
pragmatics, 141
predictability, limitation of, 3, 13
principle of superposition, 87, 130
principles (basic) of quantum mechanical method, 109, 110
probability, a posteriori determination of, 105; a priori determination of, 106; function (relative), 122; interference of, 105; statistically inferred, 105; theoretically introduced, 106
probability amplitudes, 84
probability chains, 122
probability distributions, 5, 81, 111
probability implication, 2
probability meaning, 29, 121

probability waves, 22
pseudo tertium non datur, 155
ψ-function, observational determination of the, 92; rule of the squared, 7, 84
pure case (statistical assemblages), 106

quadrangle of transformations, 62
quadratically integrable, 46, 81
quanta, Planck's theory of, 21
quantum mechanical language, 136
quantum mechanics, derivative relations of, 115

reductio ad absurdum, 156
reduction of the wave packet, 172
reference class, 95, 107
reflexive (set of basic functions), 47
relative probability function, 122
relativity, theory of, 15, 20
restricted meaning, 42; interpretation by, 141
restrictive interpretation, 33, 40, 139
reversed transformation, 58
Rosen, N., 170

scalar product, 52
Schrödinger, E., 1, 6, 21, 36, 66, 80, 95, 170
Schrödinger's first equation (time-independent), 71, 73
Schrödinger's second equation (time-dependent), 31, 70, 85
semantics, 141
semisynthetic statements, 154
simultaneity, 29
simultaneous values, 3, 119
singularities in a system of coordinates, 34
situation, physical, 81, 108
spectral decomposition, rule of, 8, 82
spectrum, 7
standard deviation, 12
stationary case, 87
statistical inference, 105
statistical laws, 1
statistically complete, 138
Strauss, M., 143
superposition, principle of, 87, 130
superposition of stationary states, 88
swarm of particles, 133
syntax, 141
synthetic statements, 150, 153; a priori, 24; fully, 154; plain, 154; semi-, 154

Tarski, A., 147, 164
tautology, 149, 153
tertium non datur, 142, 145, 149
tertium non datur, pseudo, 155
theory of relativity, 15, 20
three-valued logic, 42, 147, 150
time, direction of, 119
transforming away of anomalies, 34
triangle of transformations, 60
truth meaning, 121
truth tables, two-valued logic, 148; three-valued logic, 151
truth value, 42, 144, 148
two-canal element, 30
two-valued logic, 42, 148

unique, 46
unitary space, 55
unitary transformation, 55
unobserved objects, 17, 118

verifiability theory of meaning, 29

wave interpretation, 21, 26, 129
wave language, 146
waves, monochromatic, 7
wave number, 67
wave packet, 10; reduction of, 172
Weierstrass symbol, 46

Zilsel, E., 17

A CATALOG OF SELECTED
DOVER BOOKS
IN SCIENCE AND MATHEMATICS

CATALOG OF DOVER BOOKS

Astronomy

BURNHAM'S CELESTIAL HANDBOOK, Robert Burnham, Jr. Thorough guide to the stars beyond our solar system. Exhaustive treatment. Alphabetical by constellation: Andromeda to Cetus in Vol. 1; Chamaeleon to Orion in Vol. 2; and Pavo to Vulpecula in Vol. 3. Hundreds of illustrations. Index in Vol. 3. 2,000pp. $6^1/_8$ x $9^1/_4$.
Vol. I: 0-486-23567-X
Vol. II: 0-486-23568-8
Vol. III: 0-486-23673-0

EXPLORING THE MOON THROUGH BINOCULARS AND SMALL TELESCOPES, Ernest H. Cherrington, Jr. Informative, profusely illustrated guide to locating and identifying craters, rills, seas, mountains, other lunar features. Newly revised and updated with special section of new photos. Over 100 photos and diagrams. 240pp. $8^1/_4$ x 11. 0-486-24491-1

THE EXTRATERRESTRIAL LIFE DEBATE, 1750–1900, Michael J. Crowe. First detailed, scholarly study in English of the many ideas that developed from 1750 to 1900 regarding the existence of intelligent extraterrestrial life. Examines ideas of Kant, Herschel, Voltaire, Percival Lowell, many other scientists and thinkers. 16 illustrations. 704pp. $5^3/_8$ x $8^1/_2$. 0-486-40675-X

THEORIES OF THE WORLD FROM ANTIQUITY TO THE COPERNICAN REVOLUTION, Michael J. Crowe. Newly revised edition of an accessible, enlightening book re-creates the change from an earth-centered to a sun-centered conception of the solar system. 242pp. $5^3/_8$ x $8^1/_2$. 0-486-41444-2

ARISTARCHUS OF SAMOS: The Ancient Copernicus, Sir Thomas Heath. Heath's history of astronomy ranges from Homer and Hesiod to Aristarchus and includes quotes from numerous thinkers, compilers, and scholasticists from Thales and Anaximander through Pythagoras, Plato, Aristotle, and Heraclides. 34 figures. 448pp. $5^3/_8$ x $8^1/_2$.
0-486-43886-4

A COMPLETE MANUAL OF AMATEUR ASTRONOMY: TOOLS AND TECHNIQUES FOR ASTRONOMICAL OBSERVATIONS, P. Clay Sherrod with Thomas L. Koed. Concise, highly readable book discusses: selecting, setting up and maintaining a telescope; amateur studies of the sun; lunar topography and occultations; observations of Mars, Jupiter, Saturn, the minor planets and the stars; an introduction to photoelectric photometry; more. 1981 ed. 124 figures. 25 halftones. 37 tables. 335pp. $6^1/_2$ x $9^1/_4$. 0-486-42820-8

AMATEUR ASTRONOMER'S HANDBOOK, J. B. Sidgwick. Timeless, comprehensive coverage of telescopes, mirrors, lenses, mountings, telescope drives, micrometers, spectroscopes, more. 189 illustrations. 576pp. $5^3/_8$ x $8^1/_4$. (Available in U.S. only.)
0-486-24034-7

STAR LORE: Myths, Legends, and Facts, William Tyler Olcott. Captivating retellings of the origins and histories of ancient star groups include Pegasus, Ursa Major, Pleiades, signs of the zodiac, and other constellations. "Classic."—Sky & Telescope. 58 illustrations. 544pp. $5^3/_8$ x $8^1/_2$. 0-486-43581-4

CATALOG OF DOVER BOOKS

Chemistry

THE SCEPTICAL CHYMIST: THE CLASSIC 1661 TEXT, Robert Boyle. Boyle defines the term "element," asserting that all natural phenomena can be explained by the motion and organization of primary particles. 1911 ed. viii+232pp. $5^{3}/_{8}$ x $8^{1}/_{2}$.
0-486-42825-7

RADIOACTIVE SUBSTANCES, Marie Curie. Here is the celebrated scientist's doctoral thesis, the prelude to her receipt of the 1903 Nobel Prize. Curie discusses establishing atomic character of radioactivity found in compounds of uranium and thorium; extraction from pitchblende of polonium and radium; isolation of pure radium chloride; determination of atomic weight of radium; plus electric, photographic, luminous, heat, color effects of radioactivity. ii+94pp. $5^{3}/_{8}$ x $8^{1}/_{2}$. 0-486-42550-9

CHEMICAL MAGIC, Leonard A. Ford. Second Edition, Revised by E. Winston Grundmeier. Over 100 unusual stunts demonstrating cold fire, dust explosions, much more. Text explains scientific principles and stresses safety precautions. 128pp. $5^{3}/_{8}$ x $8^{1}/_{2}$. 0-486-67628-5

MOLECULAR THEORY OF CAPILLARITY, J. S. Rowlinson and B. Widom. History of surface phenomena offers critical and detailed examination and assessment of modern theories, focusing on statistical mechanics and application of results in mean-field approximation to model systems. 1989 edition. 352pp. $5^{3}/_{8}$ x $8^{1}/_{2}$. 0-486-42544-4

CHEMICAL AND CATALYTIC REACTION ENGINEERING, James J. Carberry. Designed to offer background for managing chemical reactions, this text examines behavior of chemical reactions and reactors; fluid-fluid and fluid-solid reaction systems; heterogeneous catalysis and catalytic kinetics; more. 1976 edition. 672pp. $6^{1}/_{8}$ x $9^{1}/_{4}$. 0-486-41736-0 $31.95

ELEMENTS OF CHEMISTRY, Antoine Lavoisier. Monumental classic by founder of modern chemistry in remarkable reprint of rare 1790 Kerr translation. A must for every student of chemistry or the history of science. 539pp. $5^{3}/_{8}$ x $8^{1}/_{2}$. 0-486-64624-6

MOLECULES AND RADIATION: An Introduction to Modern Molecular Spectroscopy. Second Edition, Jeffrey I. Steinfeld. This unified treatment introduces upper-level undergraduates and graduate students to the concepts and the methods of molecular spectroscopy and applications to quantum electronics, lasers, and related optical phenomena. 1985 edition. 512pp. $5^{3}/_{8}$ x $8^{1}/_{2}$. 0-486-44152-0

A SHORT HISTORY OF CHEMISTRY, J. R. Partington. Classic exposition explores origins of chemistry, alchemy, early medical chemistry, nature of atmosphere, theory of valency, laws and structure of atomic theory, much more. 428pp. $5^{3}/_{8}$ x $8^{1}/_{2}$. (Available in U.S. only.) 0-486-65977-1

GENERAL CHEMISTRY, Linus Pauling. Revised 3rd edition of classic first-year text by Nobel laureate. Atomic and molecular structure, quantum mechanics, statistical mechanics, thermodynamics correlated with descriptive chemistry. Problems. 992pp. $5^{3}/_{8}$ x $8^{1}/_{2}$.
0-486-65622-5

ELECTRON CORRELATION IN MOLECULES, S. Wilson. This text addresses one of theoretical chemistry's central problems. Topics include molecular electronic structure, independent electron models, electron correlation, the linked diagram theorem, and related topics. 1984 edition. 304pp. $5^{3}/_{8}$ x $8^{1}/_{2}$. 0-486-45879-2

CATALOG OF DOVER BOOKS

Engineering

DE RE METALLICA, Georgius Agricola. The famous Hoover translation of greatest treatise on technological chemistry, engineering, geology, mining of early modern times (1556). All 289 original woodcuts. 638pp. 6¾ x 11. 0-486-60006-8

FUNDAMENTALS OF ASTRODYNAMICS, Roger Bate et al. Modern approach developed by U.S. Air Force Academy. Designed as a first course. Problems, exercises. Numerous illustrations. 455pp. 5⅜ x 8½. 0-486-60061-0

DYNAMICS OF FLUIDS IN POROUS MEDIA, Jacob Bear. For advanced students of ground water hydrology, soil mechanics and physics, drainage and irrigation engineering and more. 335 illustrations. Exercises, with answers. 784pp. 6⅛ x 9¼. 0-486-65675-6

THEORY OF VISCOELASTICITY (SECOND EDITION), Richard M. Christensen. Complete consistent description of the linear theory of the viscoelastic behavior of materials. Problem-solving techniques discussed. 1982 edition. 29 figures. xiv+364pp. 6⅛ x 9¼.
0-486-42880-X

MECHANICS, J. P. Den Hartog. A classic introductory text or refresher. Hundreds of applications and design problems illuminate fundamentals of trusses, loaded beams and cables, etc. 334 answered problems. 462pp. 5⅜ x 8½. 0-486-60754-2

MECHANICAL VIBRATIONS, J. P. Den Hartog. Classic textbook offers lucid explanations and illustrative models, applying theories of vibrations to a variety of practical industrial engineering problems. Numerous figures. 233 problems, solutions. Appendix. Index. Preface. 436pp. 5⅜ x 8½. 0-486-64785-4

STRENGTH OF MATERIALS, J. P. Den Hartog. Full, clear treatment of basic material (tension, torsion, bending, etc.) plus advanced material on engineering methods, applications. 350 answered problems. 323pp. 5⅜ x 8½. 0-486-60755-0

A HISTORY OF MECHANICS, René Dugas. Monumental study of mechanical principles from antiquity to quantum mechanics. Contributions of ancient Greeks, Galileo, Leonardo, Kepler, Lagrange, many others. 671pp. 5⅜ x 8½. 0-486-65632-2

STABILITY THEORY AND ITS APPLICATIONS TO STRUCTURAL MECHANICS, Clive L. Dym. Self-contained text focuses on Koiter postbuckling analyses, with mathematical notions of stability of motion. Basing minimum energy principles for static stability upon dynamic concepts of stability of motion, it develops asymptotic buckling and postbuckling analyses from potential energy considerations, with applications to columns, plates, and arches. 1974 ed. 208pp. 5⅜ x 8½. 0-486-42541-X

BASIC ELECTRICITY, U.S. Bureau of Naval Personnel. Originally a training course; best nontechnical coverage. Topics include batteries, circuits, conductors, AC and DC, inductance and capacitance, generators, motors, transformers, amplifiers, etc. Many questions with answers. 349 illustrations. 1969 edition. 448pp. 6½ x 9¼. 0-486-20973-3

CATALOG OF DOVER BOOKS

ROCKETS, Robert Goddard. Two of the most significant publications in the history of rocketry and jet propulsion: "A Method of Reaching Extreme Altitudes" (1919) and "Liquid Propellant Rocket Development" (1936). 128pp. 5$3/8$ x 8$1/2$. 0-486-42537-1

STATISTICAL MECHANICS: PRINCIPLES AND APPLICATIONS, Terrell L. Hill. Standard text covers fundamentals of statistical mechanics, applications to fluctuation theory, imperfect gases, distribution functions, more. 448pp. 5$3/8$ x 8$1/2$. 0-486-65390-0

ENGINEERING AND TECHNOLOGY 1650–1750: ILLUSTRATIONS AND TEXTS FROM ORIGINAL SOURCES, Martin Jensen. Highly readable text with more than 200 contemporary drawings and detailed engravings of engineering projects dealing with surveying, leveling, materials, hand tools, lifting equipment, transport and erection, piling, bailing, water supply, hydraulic engineering, and more. Among the specific projects outlined-transporting a 50-ton stone to the Louvre, erecting an obelisk, building timber locks, and dredging canals. 207pp. 8$3/8$ x 11$1/4$. 0-486-42232-1

THE VARIATIONAL PRINCIPLES OF MECHANICS, Cornelius Lanczos. Graduate level coverage of calculus of variations, equations of motion, relativistic mechanics, more. First inexpensive paperbound edition of classic treatise. Index. Bibliography. 418pp. 5$3/8$ x 8$1/2$. 0-486-65067-7

PROTECTION OF ELECTRONIC CIRCUITS FROM OVERVOLTAGES, Ronald B. Standler. Five-part treatment presents practical rules and strategies for circuits designed to protect electronic systems from damage by transient overvoltages. 1989 ed. xxiv+434pp. 6$1/8$ x 9$1/4$. 0-486-42552-5

ROTARY WING AERODYNAMICS, W. Z. Stepniewski. Clear, concise text covers aerodynamic phenomena of the rotor and offers guidelines for helicopter performance evaluation. Originally prepared for NASA. 537 figures. 640pp. 6$1/8$ x 9$1/4$. 0-486-64647-5

INTRODUCTION TO SPACE DYNAMICS, William Tyrrell Thomson. Comprehensive, classic introduction to space-flight engineering for advanced undergraduate and graduate students. Includes vector algebra, kinematics, transformation of coordinates. Bibliography. Index. 352pp. 5$3/8$ x 8$1/2$. 0-486-65113-4

HISTORY OF STRENGTH OF MATERIALS, Stephen P. Timoshenko. Excellent historical survey of the strength of materials with many references to the theories of elasticity and structure. 245 figures. 452pp. 5$3/8$ x 8$1/2$. 0-486-61187-6

ANALYTICAL FRACTURE MECHANICS, David J. Unger. Self-contained text supplements standard fracture mechanics texts by focusing on analytical methods for determining crack-tip stress and strain fields. 336pp. 6$1/8$ x 9$1/4$. 0-486-41737-9

STATISTICAL MECHANICS OF ELASTICITY, J. H. Weiner. Advanced, self-contained treatment illustrates general principles and elastic behavior of solids. Part 1, based on classical mechanics, studies thermoelastic behavior of crystalline and polymeric solids. Part 2, based on quantum mechanics, focuses on interatomic force laws, behavior of solids, and thermally activated processes. For students of physics and chemistry and for polymer physicists. 1983 ed. 96 figures. 496pp. 5$3/8$ x 8$1/2$. 0-486-42260-7

CATALOG OF DOVER BOOKS

Mathematics

FUNCTIONAL ANALYSIS (Second Corrected Edition), George Bachman and Lawrence Narici. Excellent treatment of subject geared toward students with background in linear algebra, advanced calculus, physics and engineering. Text covers introduction to inner-product spaces, normed, metric spaces, and topological spaces; complete orthonormal sets, the Hahn-Banach Theorem and its consequences, and many other related subjects. 1966 ed. 544pp. 6^1/$_8$ x 9^1/$_4$. 0-486-40251-7

DIFFERENTIAL MANIFOLDS, Antoni A. Kosinski. Introductory text for advanced undergraduates and graduate students presents systematic study of the topological structure of smooth manifolds, starting with elements of theory and concluding with method of surgery. 1993 edition. 288pp. 5^3/$_8$ x 8^1/$_2$. 0-486-46244-7

VECTOR AND TENSOR ANALYSIS WITH APPLICATIONS, A. I. Borisenko and I. E. Tarapov. Concise introduction. Worked-out problems, solutions, exercises. 257pp. 5^5/$_8$ x 8^1/$_4$. 0-486-63833-2

AN INTRODUCTION TO ORDINARY DIFFERENTIAL EQUATIONS, Earl A. Coddington. A thorough and systematic first course in elementary differential equations for undergraduates in mathematics and science, with many exercises and problems (with answers). Index. 304pp. 5^3/$_8$ x 8^1/$_2$. 0-486-65942-9

FOURIER SERIES AND ORTHOGONAL FUNCTIONS, Harry F. Davis. An incisive text combining theory and practical example to introduce Fourier series, orthogonal functions and applications of the Fourier method to boundary-value problems. 570 exercises. Answers and notes. 416pp. 5^3/$_8$ x 8^1/$_2$. 0-486-65973-9

COMPUTABILITY AND UNSOLVABILITY, Martin Davis. Classic graduate-level introduction to theory of computability, usually referred to as theory of recurrent functions. New preface and appendix. 288pp. 5^3/$_8$ x 8^1/$_2$. 0-486-61471-9

AN INTRODUCTION TO MATHEMATICAL ANALYSIS, Robert A. Rankin. Dealing chiefly with functions of a single real variable, this text by a distinguished educator introduces limits, continuity, differentiability, integration, convergence of infinite series, double series, and infinite products. 1963 edition. 624pp. 5^3/$_8$ x 8^1/$_2$. 0-486-46251-X

METHODS OF NUMERICAL INTEGRATION (SECOND EDITION), Philip J. Davis and Philip Rabinowitz. Requiring only a background in calculus, this text covers approximate integration over finite and infinite intervals, error analysis, approximate integration in two or more dimensions, and automatic integration. 1984 edition. 624pp. 5^3/$_8$ x 8^1/$_2$. 0-486-45339-1

INTRODUCTION TO LINEAR ALGEBRA AND DIFFERENTIAL EQUATIONS, John W. Dettman. Excellent text covers complex numbers, determinants, orthonormal bases, Laplace transforms, much more. Exercises with solutions. Undergraduate level. 416pp. 5^3/$_8$ x 8^1/$_2$. 0-486-65191-6

RIEMANN'S ZETA FUNCTION, H. M. Edwards. Superb, high-level study of landmark 1859 publication entitled "On the Number of Primes Less Than a Given Magnitude" traces developments in mathematical theory that it inspired. xiv+315pp. 5^3/$_8$ x 8^1/$_2$.
0-486-41740-9

CATALOG OF DOVER BOOKS

CALCULUS OF VARIATIONS WITH APPLICATIONS, George M. Ewing. Applications-oriented introduction to variational theory develops insight and promotes understanding of specialized books, research papers. Suitable for advanced undergraduate/graduate students as primary, supplementary text. 352pp. $5^3/_8$ x $8^1/_2$.
0-486-64856-7

MATHEMATICIAN'S DELIGHT, W. W. Sawyer. "Recommended with confidence" by *The Times Literary Supplement*, this lively survey was written by a renowned teacher. It starts with arithmetic and algebra, gradually proceeding to trigonometry and calculus. 1943 edition. 240pp. $5^3/_8$ x $8^1/_2$.
0-486-46240-4

ADVANCED EUCLIDEAN GEOMETRY, Roger A. Johnson. This classic text explores the geometry of the triangle and the circle, concentrating on extensions of Euclidean theory, and examining in detail many relatively recent theorems. 1929 edition. 336pp. $5^3/_8$ x $8^1/_2$.
0-486-46237-4

COUNTEREXAMPLES IN ANALYSIS, Bernard R. Gelbaum and John M. H. Olmsted. These counterexamples deal mostly with the part of analysis known as "real variables." The first half covers the real number system, and the second half encompasses higher dimensions. 1962 edition. xxiv+198pp. $5^3/_8$ x $8^1/_2$.
0-486-42875-3

CATASTROPHE THEORY FOR SCIENTISTS AND ENGINEERS, Robert Gilmore. Advanced-level treatment describes mathematics of theory grounded in the work of Poincaré, R. Thom, other mathematicians. Also important applications to problems in mathematics, physics, chemistry and engineering. 1981 edition. References. 28 tables. 397 black-and-white illustrations. xvii + 666pp. $6^1/_8$ x $9^1/_4$.
0-486-67539-4

COMPLEX VARIABLES: Second Edition, Robert B. Ash and W. P. Novinger. Suitable for advanced undergraduates and graduate students, this newly revised treatment covers Cauchy theorem and its applications, analytic functions, and the prime number theorem. Numerous problems and solutions. 2004 edition. 224pp. $6^1/_2$ x $9^1/_4$.
0-486-46250-1

NUMERICAL METHODS FOR SCIENTISTS AND ENGINEERS, Richard Hamming. Classic text stresses frequency approach in coverage of algorithms, polynomial approximation, Fourier approximation, exponential approximation, other topics. Revised and enlarged 2nd edition. 721pp. $5^3/_8$ x $8^1/_2$.
0-486-65241-6

INTRODUCTION TO NUMERICAL ANALYSIS (2nd Edition), F. B. Hildebrand. Classic, fundamental treatment covers computation, approximation, interpolation, numerical differentiation and integration, other topics. 150 new problems. 669pp. $5^3/_8$ x $8^1/_2$.
0-486-65363-3

MARKOV PROCESSES AND POTENTIAL THEORY, Robert M. Blumental and Ronald K. Getoor. This graduate-level text explores the relationship between Markov processes and potential theory in terms of excessive functions, multiplicative functionals and subprocesses, additive functionals and their potentials, and dual processes. 1968 edition. 320pp. $5^3/_8$ x $8^1/_2$.
0-486-46263-3

ABSTRACT SETS AND FINITE ORDINALS: An Introduction to the Study of Set Theory, G. B. Keene. This text unites logical and philosophical aspects of set theory in a manner intelligible to mathematicians without training in formal logic and to logicians without a mathematical background. 1961 edition. 112pp. $5^3/_8$ x $8^1/_2$.
0-486-46249-8

CATALOG OF DOVER BOOKS

INTRODUCTORY REAL ANALYSIS, A.N. Kolmogorov, S. V. Fomin. Translated by Richard A. Silverman. Self-contained, evenly paced introduction to real and functional analysis. Some 350 problems. 403pp. 5⅜ x 8½. 0-486-61226-0

APPLIED ANALYSIS, Cornelius Lanczos. Classic work on analysis and design of finite processes for approximating solution of analytical problems. Algebraic equations, matrices, harmonic analysis, quadrature methods, much more. 559pp. 5⅜ x 8½. 0-486-65656-X

AN INTRODUCTION TO ALGEBRAIC STRUCTURES, Joseph Landin. Superb self-contained text covers "abstract algebra": sets and numbers, theory of groups, theory of rings, much more. Numerous well-chosen examples, exercises. 247pp. 5⅜ x 8½.
0-486-65940-2

QUALITATIVE THEORY OF DIFFERENTIAL EQUATIONS, V. V. Nemytskii and V.V. Stepanov. Classic graduate-level text by two prominent Soviet mathematicians covers classical differential equations as well as topological dynamics and ergodic theory. Bibliographies. 523pp. 5⅜ x 8½. 0-486-65954-2

THEORY OF MATRICES, Sam Perlis. Outstanding text covering rank, nonsingularity and inverses in connection with the development of canonical matrices under the relation of equivalence, and without the intervention of determinants. Includes exercises. 237pp. 5⅜ x 8½. 0-486-66810-X

INTRODUCTION TO ANALYSIS, Maxwell Rosenlicht. Unusually clear, accessible coverage of set theory, real number system, metric spaces, continuous functions, Riemann integration, multiple integrals, more. Wide range of problems. Undergraduate level. Bibliography. 254pp. 5⅜ x 8½. 0-486-65038-3

MODERN NONLINEAR EQUATIONS, Thomas L. Saaty. Emphasizes practical solution of problems; covers seven types of equations. ". . . a welcome contribution to the existing literature. . . ."—*Math Reviews.* 490pp. 5⅜ x 8½. 0-486-64232-1

MATRICES AND LINEAR ALGEBRA, Hans Schneider and George Phillip Barker. Basic textbook covers theory of matrices and its applications to systems of linear equations and related topics such as determinants, eigenvalues and differential equations. Numerous exercises. 432pp. 5⅜ x 8½. 0-486-66014-1

LINEAR ALGEBRA, Georgi E. Shilov. Determinants, linear spaces, matrix algebras, similar topics. For advanced undergraduates, graduates. Silverman translation. 387pp. 5⅜ x 8½. 0-486-63518-X

MATHEMATICAL METHODS OF GAME AND ECONOMIC THEORY: Revised Edition, Jean-Pierre Aubin. This text begins with optimization theory and convex analysis, followed by topics in game theory and mathematical economics, and concluding with an introduction to nonlinear analysis and control theory. 1982 edition. 656pp. 6⅛ x 9¼.
0-486-46265-X

SET THEORY AND LOGIC, Robert R. Stoll. Lucid introduction to unified theory of mathematical concepts. Set theory and logic seen as tools for conceptual understanding of real number system. 496pp. 5⅜ x 8¼. 0-486-63829-4

CATALOG OF DOVER BOOKS

TENSOR CALCULUS, J.L. Synge and A. Schild. Widely used introductory text covers spaces and tensors, basic operations in Riemannian space, non-Riemannian spaces, etc. 324pp. 5⅜ x 8¼. 0-486-63612-7

ORDINARY DIFFERENTIAL EQUATIONS, Morris Tenenbaum and Harry Pollard. Exhaustive survey of ordinary differential equations for undergraduates in mathematics, engineering, science. Thorough analysis of theorems. Diagrams. Bibliography. Index. 818pp. 5⅜ x 8½. 0-486-64940-7

INTEGRAL EQUATIONS, F. G. Tricomi. Authoritative, well-written treatment of extremely useful mathematical tool with wide applications. Volterra Equations, Fredholm Equations, much more. Advanced undergraduate to graduate level. Exercises. Bibliography. 238pp. 5⅜ x 8½. 0-486-64828-1

FOURIER SERIES, Georgi P. Tolstov. Translated by Richard A. Silverman. A valuable addition to the literature on the subject, moving clearly from subject to subject and theorem to theorem. 107 problems, answers. 336pp. 5⅜ x 8½. 0-486-63317-9

INTRODUCTION TO MATHEMATICAL THINKING, Friedrich Waismann. Examinations of arithmetic, geometry, and theory of integers; rational and natural numbers; complete induction; limit and point of accumulation; remarkable curves; complex and hypercomplex numbers, more. 1959 ed. 27 figures. xii+260pp. 5⅜ x 8½. 0-486-42804-8

THE RADON TRANSFORM AND SOME OF ITS APPLICATIONS, Stanley R. Deans. Of value to mathematicians, physicists, and engineers, this excellent introduction covers both theory and applications, including a rich array of examples and literature. Revised and updated by the author. 1993 edition. 304pp. 6⅛ x 9¼. 0-486-46241-2

CALCULUS OF VARIATIONS, Robert Weinstock. Basic introduction covering isoperimetric problems, theory of elasticity, quantum mechanics, electrostatics, etc. Exercises throughout. 326pp. 5⅜ x 8½. 0-486-63069-2

THE CONTINUUM: A CRITICAL EXAMINATION OF THE FOUNDATION OF ANALYSIS, Hermann Weyl. Classic of 20th-century foundational research deals with the conceptual problem posed by the continuum. 156pp. 5⅜ x 8½. 0-486-67982-9

CHALLENGING MATHEMATICAL PROBLEMS WITH ELEMENTARY SOLUTIONS, A. M. Yaglom and I. M. Yaglom. Over 170 challenging problems on probability theory, combinatorial analysis, points and lines, topology, convex polygons, many other topics. Solutions. Total of 445pp. 5⅜ x 8½. Two-vol. set.
Vol. I: 0-486-65536-9   Vol. II: 0-486-65537-7

INTRODUCTION TO PARTIAL DIFFERENTIAL EQUATIONS WITH APPLICATIONS, E. C. Zachmanoglou and Dale W. Thoe. Essentials of partial differential equations applied to common problems in engineering and the physical sciences. Problems and answers. 416pp. 5⅜ x 8½. 0-486-65251-3

STOCHASTIC PROCESSES AND FILTERING THEORY, Andrew H. Jazwinski. This unified treatment presents material previously available only in journals, and in terms accessible to engineering students. Although theory is emphasized, it discusses numerous practical applications as well. 1970 edition. 400pp. 5⅜ x 8½. 0-486-46274-9

CATALOG OF DOVER BOOKS

Math—Decision Theory, Statistics, Probability

INTRODUCTION TO PROBABILITY, John E. Freund. Featured topics include permutations and factorials, probabilities and odds, frequency interpretation, mathematical expectation, decision-making, postulates of probability, rule of elimination, much more. Exercises with some solutions. Summary. 1973 edition. 247pp. 5³/₈ x 8¹/₂.
0-486-67549-1

STATISTICAL AND INDUCTIVE PROBABILITIES, Hugues Leblanc. This treatment addresses a decades-old dispute among probability theorists, asserting that both statistical and inductive probabilities may be treated as sentence-theoretic measurements, and that the latter qualify as estimates of the former. 1962 edition. 160pp. 5³/₈ x 8¹/₂.
0-486-44980-7

APPLIED MULTIVARIATE ANALYSIS: Using Bayesian and Frequentist Methods of Inference, Second Edition, S. James Press. This two-part treatment deals with foundations as well as models and applications. Topics include continuous multivariate distributions; regression and analysis of variance; factor analysis and latent structure analysis; and structuring multivariate populations. 1982 edition. 692pp. 5³/₈ x 8¹/₂.
0-486-44236-5

LINEAR PROGRAMMING AND ECONOMIC ANALYSIS, Robert Dorfman, Paul A. Samuelson and Robert M. Solow. First comprehensive treatment of linear programming in standard economic analysis. Game theory, modern welfare economics, Leontief input-output, more. 525pp. 5³/₈ x 8¹/₂.
0-486-65491-5

PROBABILITY: AN INTRODUCTION, Samuel Goldberg. Excellent basic text covers set theory, probability theory for finite sample spaces, binomial theorem, much more. 360 problems. Bibliographies. 322pp. 5³/₈ x 8¹/₂.
0-486-65252-1

GAMES AND DECISIONS: INTRODUCTION AND CRITICAL SURVEY, R. Duncan Luce and Howard Raiffa. Superb nontechnical introduction to game theory, primarily applied to social sciences. Utility theory, zero-sum games, n-person games, decision-making, much more. Bibliography. 509pp. 5³/₈ x 8¹/₂.
0-486-65943-7

INTRODUCTION TO THE THEORY OF GAMES, J. C. C. McKinsey. This comprehensive overview of the mathematical theory of games illustrates applications to situations involving conflicts of interest, including economic, social, political, and military contexts. Appropriate for advanced undergraduate and graduate courses; advanced calculus a prerequisite. 1952 ed. x+372pp. 5³/₈ x 8¹/₂.
0-486-42811-7

FIFTY CHALLENGING PROBLEMS IN PROBABILITY WITH SOLUTIONS, Frederick Mosteller. Remarkable puzzlers, graded in difficulty, illustrate elementary and advanced aspects of probability. Detailed solutions. 88pp. 5³/₈ x 8¹/₂.
0-486-65355-2

PROBABILITY THEORY: A CONCISE COURSE, Y. A. Rozanov. Highly readable, self-contained introduction covers combination of events, dependent events, Bernoulli trials, etc. 148pp. 5³/₈ x 8¹/₄.
0-486-63544-9

THE STATISTICAL ANALYSIS OF EXPERIMENTAL DATA, John Mandel. First half of book presents fundamental mathematical definitions, concepts and facts while remaining half deals with statistics primarily as an interpretive tool. Well-written text, numerous worked examples with step-by-step presentation. Includes 116 tables. 448pp. 5³/₈ x 8¹/₂.
0-486-64666-1

CATALOG OF DOVER BOOKS

Math—Geometry and Topology

ELEMENTARY CONCEPTS OF TOPOLOGY, Paul Alexandroff. Elegant, intuitive approach to topology from set-theoretic topology to Betti groups; how concepts of topology are useful in math and physics. 25 figures. 57pp. 5⅜ x 8½. 0-486-60747-X

A LONG WAY FROM EUCLID, Constance Reid. Lively guide by a prominent historian focuses on the role of Euclid's Elements in subsequent mathematical developments. Elementary algebra and plane geometry are sole prerequisites. 80 drawings. 1963 edition. 304pp. 5⅜ x 8½. 0-486-43613-6

EXPERIMENTS IN TOPOLOGY, Stephen Barr. Classic, lively explanation of one of the byways of mathematics. Klein bottles, Moebius strips, projective planes, map coloring, problem of the Koenigsberg bridges, much more, described with clarity and wit. 43 figures. 210pp. 5⅜ x 8½. 0-486-25933-1

THE GEOMETRY OF RENÉ DESCARTES, René Descartes. The great work founded analytical geometry. Original French text, Descartes's own diagrams, together with definitive Smith-Latham translation. 244pp. 5⅜ x 8½. 0-486-60068-8

EUCLIDEAN GEOMETRY AND TRANSFORMATIONS, Clayton W. Dodge. This introduction to Euclidean geometry emphasizes transformations, particularly isometries and similarities. Suitable for undergraduate courses, it includes numerous examples, many with detailed answers. 1972 ed. viii+296pp. 6⅛ x 9¼. 0-486-43476-1

EXCURSIONS IN GEOMETRY, C. Stanley Ogilvy. A straightedge, compass, and a little thought are all that's needed to discover the intellectual excitement of geometry. Harmonic division and Apollonian circles, inversive geometry, hexlet, Golden Section, more. 132 illustrations. 192pp. 5⅜ x 8½. 0-486-26530-7

THE THIRTEEN BOOKS OF EUCLID'S ELEMENTS, translated with introduction and commentary by Sir Thomas L. Heath. Definitive edition. Textual and linguistic notes, mathematical analysis. 2,500 years of critical commentary. Unabridged. 1,414pp. 5⅜ x 8½. Three-vol. set.
    Vol. I: 0-486-60088-2   Vol. II: 0-486-60089-0   Vol. III: 0-486-60090-4

SPACE AND GEOMETRY: IN THE LIGHT OF PHYSIOLOGICAL, PSYCHOLOGICAL AND PHYSICAL INQUIRY, Ernst Mach. Three essays by an eminent philosopher and scientist explore the nature, origin, and development of our concepts of space, with a distinctness and precision suitable for undergraduate students and other readers. 1906 ed. vi+148pp. 5⅜ x 8½. 0-486-43909-7

GEOMETRY OF COMPLEX NUMBERS, Hans Schwerdtfeger. Illuminating, widely praised book on analytic geometry of circles, the Moebius transformation, and two-dimensional non-Euclidean geometries. 200pp. 5⅜ x 8¼. 0-486-63830-8

DIFFERENTIAL GEOMETRY, Heinrich W. Guggenheimer. Local differential geometry as an application of advanced calculus and linear algebra. Curvature, transformation groups, surfaces, more. Exercises. 62 figures. 378pp. 5⅜ x 8½. 0-486-63433-7

CATALOG OF DOVER BOOKS

History of Math

THE WORKS OF ARCHIMEDES, Archimedes (T. L. Heath, ed.). Topics include the famous problems of the ratio of the areas of a cylinder and an inscribed sphere; the measurement of a circle; the properties of conoids, spheroids, and spirals; and the quadrature of the parabola. Informative introduction. clxxxvi+326pp. 5⅜ x 8½. 0-486-42084-1

A SHORT ACCOUNT OF THE HISTORY OF MATHEMATICS, W. W. Rouse Ball. One of clearest, most authoritative surveys from the Egyptians and Phoenicians through 19th-century figures such as Grassman, Galois, Riemann. Fourth edition. 522pp. 5⅜ x 8½. 0-486-20630-0

THE HISTORY OF THE CALCULUS AND ITS CONCEPTUAL DEVELOPMENT, Carl B. Boyer. Origins in antiquity, medieval contributions, work of Newton, Leibniz, rigorous formulation. Treatment is verbal. 346pp. 5⅜ x 8½. 0-486-60509-4

THE HISTORICAL ROOTS OF ELEMENTARY MATHEMATICS, Lucas N. H. Bunt, Phillip S. Jones, and Jack D. Bedient. Fundamental underpinnings of modern arithmetic, algebra, geometry and number systems derived from ancient civilizations. 320pp. 5⅜ x 8½. 0-486-25563-8

THE HISTORY OF THE CALCULUS AND ITS CONCEPTUAL DEVELOPMENT, Carl B. Boyer. Fluent description of the development of both the integral and differential calculus—its early beginnings in antiquity, medieval contributions, and a consideration of Newton and Leibniz. 368pp. 5⅜ x 8½. 0-486-60509-4

GAMES, GODS & GAMBLING: A HISTORY OF PROBABILITY AND STATISTICAL IDEAS, F. N. David. Episodes from the lives of Galileo, Fermat, Pascal, and others illustrate this fascinating account of the roots of mathematics. Features thought-provoking references to classics, archaeology, biography, poetry. 1962 edition. 304pp. 5⅜ x 8½. (Available in U.S. only.) 0-486-40023-9

OF MEN AND NUMBERS: THE STORY OF THE GREAT MATHEMATICIANS, Jane Muir. Fascinating accounts of the lives and accomplishments of history's greatest mathematical minds—Pythagoras, Descartes, Euler, Pascal, Cantor, many more. Anecdotal, illuminating. 30 diagrams. Bibliography. 256pp. 5⅜ x 8½. 0-486-28973-7

HISTORY OF MATHEMATICS, David E. Smith. Nontechnical survey from ancient Greece and Orient to late 19th century; evolution of arithmetic, geometry, trigonometry, calculating devices, algebra, the calculus. 362 illustrations. 1,355pp. 5⅜ x 8½. Two-vol. set. Vol. I: 0-486-20429-4   Vol. II: 0-486-20430-8

A CONCISE HISTORY OF MATHEMATICS, Dirk J. Struik. The best brief history of mathematics. Stresses origins and covers every major figure from ancient Near East to 19th century. 41 illustrations. 195pp. 5⅜ x 8½. 0-486-60255-9

CATALOG OF DOVER BOOKS

Physics

OPTICAL RESONANCE AND TWO-LEVEL ATOMS, L. Allen and J. H. Eberly. Clear, comprehensive introduction to basic principles behind all quantum optical resonance phenomena. 53 illustrations. Preface. Index. 256pp. 5⅜ x 8½. 0-486-65533-4

QUANTUM THEORY, David Bohm. This advanced undergraduate-level text presents the quantum theory in terms of qualitative and imaginative concepts, followed by specific applications worked out in mathematical detail. Preface. Index. 655pp. 5⅜ x 8½.
0-486-65969-0

ATOMIC PHYSICS (8th EDITION), Max Born. Nobel laureate's lucid treatment of kinetic theory of gases, elementary particles, nuclear atom, wave-corpuscles, atomic structure and spectral lines, much more. Over 40 appendices, bibliography. 495pp. 5⅜ x 8½.
0-486-65984-4

A SOPHISTICATE'S PRIMER OF RELATIVITY, P. W. Bridgman. Geared toward readers already acquainted with special relativity, this book transcends the view of theory as a working tool to answer natural questions: What is a frame of reference? What is a "law of nature"? What is the role of the "observer"? Extensive treatment, written in terms accessible to those without a scientific background. 1983 ed. xlviii+172pp. 5⅜ x 8½.
0-486-42549-5

AN INTRODUCTION TO HAMILTONIAN OPTICS, H. A. Buchdahl. Detailed account of the Hamiltonian treatment of aberration theory in geometrical optics. Many classes of optical systems defined in terms of the symmetries they possess. Problems with detailed solutions. 1970 edition. xv + 360pp. 5⅜ x 8½. 0-486-67597-1

PRIMER OF QUANTUM MECHANICS, Marvin Chester. Introductory text examines the classical quantum bead on a track: its state and representations; operator eigenvalues; harmonic oscillator and bound bead in a symmetric force field; and bead in a spherical shell. Other topics include spin, matrices, and the structure of quantum mechanics; the simplest atom; indistinguishable particles; and stationary-state perturbation theory. 1992 ed. xiv+314pp. 6⅛ x 9¼. 0-486-42878-8

LECTURES ON QUANTUM MECHANICS, Paul A. M. Dirac. Four concise, brilliant lectures on mathematical methods in quantum mechanics from Nobel Prize-winning quantum pioneer build on idea of visualizing quantum theory through the use of classical mechanics. 96pp. 5⅜ x 8½. 0-486-41713-1

THIRTY YEARS THAT SHOOK PHYSICS: THE STORY OF QUANTUM THEORY, George Gamow. Lucid, accessible introduction to influential theory of energy and matter. Careful explanations of Dirac's anti-particles, Bohr's model of the atom, much more. 12 plates. Numerous drawings. 240pp. 5⅜ x 8½. 0-486-24895-X

ELECTRONIC STRUCTURE AND THE PROPERTIES OF SOLIDS: THE PHYSICS OF THE CHEMICAL BOND, Walter A. Harrison. Innovative text offers basic understanding of the electronic structure of covalent and ionic solids, simple metals, transition metals and their compounds. Problems. 1980 edition. 582pp. 6⅛ x 9¼.
0-486-66021-4

CATALOG OF DOVER BOOKS

HYDRODYNAMIC AND HYDROMAGNETIC STABILITY, S. Chandrasekhar. Lucid examination of the Rayleigh-Benard problem; clear coverage of the theory of instabilities causing convection. 704pp. 5⅜ x 8¼. 0-486-64071-X

INVESTIGATIONS ON THE THEORY OF THE BROWNIAN MOVEMENT, Albert Einstein. Five papers (1905–8) investigating dynamics of Brownian motion and evolving elementary theory. Notes by R. Fürth. 122pp. 5⅜ x 8½. 0-486-60304-0

THE PHYSICS OF WAVES, William C. Elmore and Mark A. Heald. Unique overview of classical wave theory. Acoustics, optics, electromagnetic radiation, more. Ideal as classroom text or for self-study. Problems. 477pp. 5⅜ x 8½. 0-486-64926-1

GRAVITY, George Gamow. Distinguished physicist and teacher takes reader-friendly look at three scientists whose work unlocked many of the mysteries behind the laws of physics: Galileo, Newton, and Einstein. Most of the book focuses on Newton's ideas, with a concluding chapter on post-Einsteinian speculations concerning the relationship between gravity and other physical phenomena. 160pp. 5⅜ x 8½. 0-486-42563-0

PHYSICAL PRINCIPLES OF THE QUANTUM THEORY, Werner Heisenberg. Nobel Laureate discusses quantum theory, uncertainty, wave mechanics, work of Dirac, Schroedinger, Compton, Wilson, Einstein, etc. 184pp. 5⅜ x 8½. 0-486-60113-7

ATOMIC SPECTRA AND ATOMIC STRUCTURE, Gerhard Herzberg. One of best introductions; especially for specialist in other fields. Treatment is physical rather than mathematical. 80 illustrations. 257pp. 5⅜ x 8½. 0-486-60115-3

AN INTRODUCTION TO STATISTICAL THERMODYNAMICS, Terrell L. Hill. Excellent basic text offers wide-ranging coverage of quantum statistical mechanics, systems of interacting molecules, quantum statistics, more. 523pp. 5⅜ x 8½. 0-486-65242-4

THEORETICAL PHYSICS, Georg Joos, with Ira M. Freeman. Classic overview covers essential math, mechanics, electromagnetic theory, thermodynamics, quantum mechanics, nuclear physics, other topics. First paperback edition. xxiii + 885pp. 5⅜ x 8½.
0-486-65227-0

PROBLEMS AND SOLUTIONS IN QUANTUM CHEMISTRY AND PHYSICS, Charles S. Johnson, Jr. and Lee G. Pedersen. Unusually varied problems, detailed solutions in coverage of quantum mechanics, wave mechanics, angular momentum, molecular spectroscopy, more. 280 problems plus 139 supplementary exercises. 430pp. 6½ x 9¼.
0-486-65236-X

THEORETICAL SOLID STATE PHYSICS, Vol. 1: Perfect Lattices in Equilibrium; Vol. II: Non-Equilibrium and Disorder, William Jones and Norman H. March. Monumental reference work covers fundamental theory of equilibrium properties of perfect crystalline solids, non-equilibrium properties, defects and disordered systems. Appendices. Problems. Preface. Diagrams. Index. Bibliography. Total of 1,301pp. 5⅜ x 8½. Two volumes. Vol. I: 0-486-65015-4 Vol. II: 0-486-65016-2

WHAT IS RELATIVITY? L. D. Landau and G. B. Rumer. Written by a Nobel Prize physicist and his distinguished colleague, this compelling book explains the special theory of relativity to readers with no scientific background, using such familiar objects as trains, rulers, and clocks. 1960 ed. vi+72pp. 5⅜ x 8½. 0-486-42806-0

CATALOG OF DOVER BOOKS

A TREATISE ON ELECTRICITY AND MAGNETISM, James Clerk Maxwell. Important foundation work of modern physics. Brings to final form Maxwell's theory of electromagnetism and rigorously derives his general equations of field theory. 1,084pp. 5^3/$_8$ x 8^1/$_2$. Two-vol. set. Vol. I: 0-486-60636-8 Vol. II: 0-486-60637-6

MATHEMATICS FOR PHYSICISTS, Philippe Dennery and Andre Krzywicki. Superb text provides math needed to understand today's more advanced topics in physics and engineering. Theory of functions of a complex variable, linear vector spaces, much more. Problems. 1967 edition. 400pp. 6^1/$_2$ x 9^1/$_4$. 0-486-69193-4

INTRODUCTION TO QUANTUM MECHANICS WITH APPLICATIONS TO CHEMISTRY, Linus Pauling & E. Bright Wilson, Jr. Classic undergraduate text by Nobel Prize winner applies quantum mechanics to chemical and physical problems. Numerous tables and figures enhance the text. Chapter bibliographies. Appendices. Index. 468pp. 5^3/$_8$ x 8^1/$_2$. 0-486-64871-0

METHODS OF THERMODYNAMICS, Howard Reiss. Outstanding text focuses on physical technique of thermodynamics, typical problem areas of understanding, and significance and use of thermodynamic potential. 1965 edition. 238pp. 5^3/$_8$ x 8^1/$_2$. 0-486-69445-3

THE ELECTROMAGNETIC FIELD, Albert Shadowitz. Comprehensive under- graduate text covers basics of electric and magnetic fields, builds up to electromagnetic theory. Also related topics, including relativity. Over 900 problems. 768pp. 5^3/$_8$ x 8^1/$_2$. 0-486-65660-8

GREAT EXPERIMENTS IN PHYSICS: FIRSTHAND ACCOUNTS FROM GALILEO TO EINSTEIN, Morris H. Shamos (ed.). 25 crucial discoveries: Newton's laws of motion, Chadwick's study of the neutron, Hertz on electromagnetic waves, more. Original accounts clearly annotated. 370pp. 5^3/$_8$ x 8^1/$_2$. 0-486-25346-5

EINSTEIN'S LEGACY, Julian Schwinger. A Nobel Laureate relates fascinating story of Einstein and development of relativity theory in well-illustrated, nontechnical volume. Subjects include meaning of time, paradoxes of space travel, gravity and its effect on light, non-Euclidean geometry and curving of space-time, impact of radio astronomy and space-age discoveries, and more. 189 b/w illustrations. xiv+250pp. 8^3/$_8$ x 9^1/$_4$. 0-486-41974-6

THE VARIATIONAL PRINCIPLES OF MECHANICS, Cornelius Lanczos. Philosophic, less formalistic approach to analytical mechanics offers model of clear, scholarly exposition at graduate level with coverage of basics, calculus of variations, principle of virtual work, equations of motion, more. 418pp. 5^3/$_8$ x 8^1/$_2$. 0-486-65067-7

Paperbound unless otherwise indicated. Available at your book dealer, online at www.doverpublications.com, or by writing to Dept. GI, Dover Publications, Inc., 31 East 2nd Street, Mineola, NY 11501. For current price information or for free catalogues (please indicate field of interest), write to Dover Publications or log on to www.doverpublications.com and see every Dover book in print. Dover publishes more than 400 books each year on science, elementary and advanced mathematics, biology, music, art, literary history, social sciences, and other areas.